Bio-Kernel Machines and Applications

Bio-Kernel
Machines and
Applications

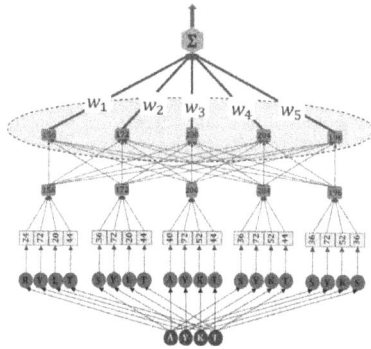

Zheng Rong Yang
University of Exeter, UK

 World Scientific

NEW JERSEY · LONDON · SINGAPORE · BEIJING · SHANGHAI · HONG KONG · TAIPEI · CHENNAI · TOKYO

Published by

World Scientific Publishing Co. Pte. Ltd.

5 Toh Tuck Link, Singapore 596224

USA office: 27 Warren Street, Suite 401-402, Hackensack, NJ 07601

UK office: 57 Shelton Street, Covent Garden, London WC2H 9HE

Library of Congress Cataloging-in-Publication Data

Names: Yang, Zheng Rong author.

Title: Bio-kernel machines and applications / Zheng Rong Yang, University of Exeter, UK.

Description: New Jersey : World Scientific, [2024] | Includes bibliographical references and index.

Identifiers: LCCN 2024000266 | ISBN 9789811287336 (hardcover) |
   ISBN 9789811287343 (ebook for institutions) | ISBN 9789811287350 (ebook for individuals)

Subjects: LCSH: Bioinformatics. | Kernel functions. | Machine learning. |
   Pattern recognition systems.

Classification: LCC QH324.25 .Y259 2024 | DDC 570.285--dc23/eng/20240205

LC record available at https://lccn.loc.gov/2024000266

British Library Cataloguing-in-Publication Data

A catalogue record for this book is available from the British Library.

For any available supplementary material, please visit
https://www.worldscientific.com/worldscibooks/10.1142/13704#t=suppl

Desk Editor: Vanessa Quek ZhiQin

Typeset by Stallion Press
Email: enquiries@stallionpress.com

To Caizhen Wang & Zihua Yang

Preface

In this big-data era, developing and employing machine learning approaches for discovering and analysing patterns hidden in data has been widely exercised with successes in almost all real-world applications, including biology. Protein peptide research using machine learning approaches has also been widely practiced, leading to many successful pattern discoveries. The resulting well-learned models can organise inherent genotypic-phenotypic logic into patterns, which supports further biological research. For instance, HIV inhibitor design may not be possible without the successful research on HIV protease cleavage peptides. However, peptides are non-numerical data and each residue of a peptide is occupied by an amino acid which is an organic molecule. Therefore, analysing peptide data requires a different strategy. The first thing is to convert non-numerical peptides into numerical data so that a machine learning algorithm can be employed to analyse the data. This process is referred to as coding or feature extraction. The basic requirement of a successful and meaningful coding process is its accuracy in biology. A coding process departing from the biological constrain may lead to a converted data set with low accuracy by which the outcome of a model, if built, cannot be well-interpreted. This is because one pair of amino acids may have different mutual biological interaction from another pair of amino acids. In other words, the interaction or the relationship between amino acids is not a constant across all amino acids. One amino acid may tend to have a closer relationship with the other amino acid when they have closer biochemical characteristic. A pair of amino acids which share similar biochemical property may have a greater likelihood to mutate to each other in evolution but doing so would not change the protein structure significantly. There have been many new ideas and algorithms for coding amino acids that have worked and produced positive results. For instance, the orthogonal, physical-chemical descriptor and pseudo coding approaches have been successfully developed and employed in various peptide data analysis projects. As another new contribution, the bio-kernel machine is also developed for this purpose with success. The unique feature of this new approach is that it tends to have more biology-sound treatments for amino acids. Rather than treating each amino acid as an individual and independent physical object for coding, this new approach concerns more on what amino acids are in biology. Therefore, it treats amino acids as the mutually-related components in a system. Technically, it employs a mutation matrix to quantify the relationship between amino acids. Since its development, the bio-kernel machine has been applied to many peptide data analysis projects. It has also been evolved to several advanced versions which will be introduced in this book.

This book is written based on the author's research work in collaboration with his colleagues in the University of Exeter and other universities from the United Kingdom, as well as international institutions. The objective of this book is to provide the working principle of bio-kernel machines for researchers who work in this area and for them to gather new ideas in this area. The book is written in a sequential order from the initial version of the bio-kernel machine to the advanced versions. Therefore, the first few chapters of this book may provide basis to the following chapters.

Finally, the author must make a sincere acknowledgement to his family, especially his wife and his daughter. Without their sustained support and encouragement, the author may not have the strength to complete this book. The author also wishes his elderly parents well, as he feels guilty of not being able to have more time to accompany them when writing this book. When pre-reviewing the manuscript of this book, the author reminds himself his middle school teacher, Mr Shao Guangyu (邵光宇先生). He used to have many one-to-one chats with the author when the author was studying in the middle school. Particularly, the author still remembers the words of Mr Shao *"rigorous is the attitude and is your reputation too."* Coincidently, the post-doctoral supervisor of the author, Professor Robert Harrison, also had the same words many years later when the author was working with him. The author's gratefulness is thus devoted to them too.

October 28, 2023
Edinburgh

Contents

Chapter 1

Introduction

As a key component in artificial intelligence, machine learning plays an increasingly important role in real-world data analysis from pattern recognition to intelligent inference, from data structure reservation to knowledge discovery, from finance to media, from chemistry to physics, from medicine to biology, etc. The significance of machine learning in many real-world applications is not only because of its enhanced generalization capability compared to traditional univariate statistical test approaches but also because of its robustness as well as its inference power (Savage, 2017; Butler *et al.*, 2018; Wong and Yip, 2018; Reardon, 2019; Hein, 2021; Sammut *et al.*, 2022; Bures and Larrosa, 2023; Lustosa and Milo, 2023). Machine learning has also gradually changed its role in real-world applications, i.e., from a tool supplier or a toolbox maker to a collaborator and a participator in many subjects, including biological pattern discovery research.

Biology may be one of the most challenging areas for employing machine learning approaches. Biological data pattern recognition covers a very wide range of subjects from disease prevention to diagnosis/prognosis, from gene expression analysis to protein structure discovery, from metabolite function study to protein function research, from molecular interaction investigation to cellular network reconstruction, etc. Analyzing protein sequences is one of the subjects and has been practiced for more than a century, but research in this area is still challenging due to several reasons. *First*, the quantity of sequence data nowadays is several magnitudes greater than a century ago. In addition, computing facilities still need more

1

improvement, although there is no doubt that it is hard to imagine how Saul Needleman and Christian Wunsch would revise their sequence alignment algorithms if they could use today's computer and today's data or how Margret Dayhoff and her team would revise their amino acid mutation matrix if she and her colleagues could use today's computing facility. *Second*, research into the complexity of biological data is much deeper than before. When some unknowns become knowns, some knowns turn back to unknowns. This thus brings more challenges into research and this is why multi-discipline research teams have emerged around the world to tackle many hard and tough biological research subjects. In such circumstances, machine learning has been a unique, powerful, and non-replaceable participant in research.

The research on protein functional sites based on the data of short-sequence segments, which are referred to as peptides with a length up to a dozen residues, is one of the key areas in protein sequence analysis. Post-translational modification pattern discovery and protease cleavage pattern discovery are two typical research topics in this area. One of the important outcomes of the research is the drug design, discovery, and development (Cottier *et al.*, 2006; Moelleken *et al.*, 2017; Meng *et al.*, 2021; Kaiser *et al.*, 2022). For instance, the design and development of HIV inhibitor cannot be carried out alone without the research of the pattern of HIV protease cleavage sites in peptides (Beck *et al.*, 2000; Cofe *et al.*, 2001; Gatanaga *et al.*, 2002). There have been many ongoing efforts to improve algorithm efficiency, quality, accuracy, and robustness in peptide pattern discovery and research.

To analyze peptides for protein functional site pattern discovery and research, the first issue is how to convert a non-numerical peptide space to a numerical space, in which conventional machine learning algorithms can be used to analyze the data. This process is called coding or feature extraction in machine learning. The most commonly employed coding approaches for peptide data analysis include the orthogonal coding approach (Cai and Chou, 1998), the physio-chemical descriptors coding approach (Tong *et al.*, 2008; Liang *et al.*, 2009; Xie *et al.*, 2010; Yang *et al.*, 2010; van Westen *et al.*, 2013), and the pseudo-coding approach (Chou, 2011). The orthogonal coding approach is the closest to the earliest sequence alignment algorithms, such as the Sellers algorithm (Sellers, 1974) and the Needleman–Wunsch algorithm (Needleman and Wunsch, 1970), where the edit distance is used to measure the similarity or distance between two residues from two sequences under alignment. The physio-chemical descriptors are generally developed for drug design, such as quantitative structure–activity relationship (QSAR) research (Wisniowska *et al.*, 2015; Kelleci and Karaduman, 2023; Mousavi and Sajjadi, 2023). Although the descriptors can describe the biochemical property of amino acids, they have little information regarding the probability or the likelihood of how one amino acid is mutated to the other during evolution. Therefore, their coding power is limited as well. The pseudo-coding approach is certainly statistically sound. It examines the occurrence of amino acids and the coupling effect between residues in peptides. The pattern that is explored this way is used to code amino acids in peptides. However, the method was developed based on the statistics of the frequency of amino acids as well as the frequency of neighbor amino acids. Therefore, it cannot explore the information of how likely an amino acid is mutated to the other in peptides. All these coding approaches have the same property, i.e., transferring a peptide space to a numerical space with one-to-one correspondence as shown in the following:

$$\phi : \mathbf{s} \mapsto \mathbf{x} \tag{1.1}$$

where $\mathcal{A}$ is a set of amino acids, $\mathbf{s} \in \mathcal{A}^R$ is a peptide with R residues, and $\mathbf{x} = \phi(\mathbf{s}) \in \mathcal{R}^d$ is a numeric vector in a d-dimensional space. Note that $d > R$.

In contrast, the bio-kernel machine takes the amino acid mutation probabilities or scores into the consideration in peptide pattern discovery and research. The most important feature of the bio-kernel machine is its utilization of the kernel approach. With the use of a kernel function, the mapping from a peptide space to a numerical kernel space is fully supported by the internal biological content. The kernel mapping principle of the bio-kernel machine is shown as follows:

$$k = \mathcal{K}(\mathbf{s}, \mathcal{H}) \tag{1.2}$$

where $\mathcal{K}$ is a kernel function, $\mathcal{H} = (\boldsymbol{\nu}_1, \boldsymbol{\nu}_2, \ldots, \boldsymbol{\nu}_H) \in \mathcal{R}^{H \times R}$ is a set of H hypothetical kernel peptides, and $k \in \mathcal{R}^H$ is an H-dimensional numerical vector corresponding to a peptide $\mathbf{s} \in \mathcal{A}^R$. This improvement thus makes a model constructed using the bio-kernel machine more biologically sound. Its promising property is the capability of interpreting the prediction of a peptide. For instance, why a peptide is predicted as functional and what the significance a prediction is can be interpreted.

This book starts from a very simple and naïve introduction to linear algebra in Chapter 2, aiming to establish its later use in the book. Chapter 3 introduces some traditional or classical kernel machines. In this chapter, the kernel principles and most relevant kernel machines for the introduction of bio-kernel machines are presented. Importantly, the kernel trick is introduced as well because it plays a key role in kernel machine development. Chapter 4 introduces several kernel machines which are developed for whole-sequence analysis, such as string kernels, wildcard kernels, and mismatch kernels. The introduction of these kernels can show how a bio-kernel machine is developed based on these whole-sequence kernel machines. Chapter 5 introduces how mutation matrices are developed because they are a major component of bio-kernel machines. The chapter will introduce the Dayhoff mutation matrix, which is the first mutation matrix developed to quantify how likely an amino acid is to mutate to the other amino acids and is the fundamental cornerstone for later sequence alignment algorithm development. Chapter 6 introduces the original bio-kernel machines, including the least-squares bio-kernel machine and the Fisher discriminant bio-kernel machine. Chapter 7 introduces several advanced bio-kernel machines, including the orthogonal bio-kernel machine, the Bayesian bio-kernel machine, the intelligent bio-kernel machine, and the deep bio-kernel machine. These advanced bio-kernel machines have different strengths compared to the original bio-kernel machines. For instance, they can generate parsimonious bio-kernel machine models and improve the generalization capability of the bio-kernel machine models. Chapter 8 introduces how fusion technology is employed in dealing with the uncertainty of mutation matrix selection when constructing a bio-kernel machine model for pattern discovery and research for a peptide dataset. Importantly, a data-driven mutation matrix estimation approach is also introduced. It means that a specific mutation matrix is estimated from a given dataset when the dataset is targeted to generate a predictive model for peptide functionality prediction. To enhance the accuracy and robustness of the data-driven mutation matrix estimation, the Monte Carlo simulation approach is used. Chapter 9 presents the kernelized principal component analysis approach and the kernelized self-organizing map approach that introduce the visualization capability and supervised learning capability for bio-kernel machines. Chapter 10 concludes on the issues of future research directions of bio-kernel machines.

Chapter 2

Basic Algebra

Linear algebra is the foundation of most machine learning approaches. This chapter briefly introduces basic linear algebra which will be used in later chapters for bio-kernel machines.

2.1 Vector and matrix

A vector is normally defined as a collection of a number of data points, which have the same property, that represent an object in a multi-dimensional space and take part in calculations. For instance, a vector can be used to collect the expressions of a gene sample from a number of patients who have the same medical background. Suppose two groups of patients have participated in a study and one group has developed cancer, while the other group has not developed cancer. If two vectors have been used to collect the expressions of a suspected cancerous gene sampled from these two groups of patients, the two vectors can be used in a statistical analysis such as the t test to investigate whether this gene is a cancerous gene.

Although in most situations a vector refers to a collection of numerical data, it is often noted that a vector can be composed of non-numerical data. For instance, a vector can be composed of a number of amino acids that represent a protein peptide. In the following text, a vector is composed of numerical data if no specific indication is given.

A vector is usually expressed as a column in linear algebra. A vector of n entries is denoted as shown in the following, where x_i is the ith entry of the $\mathbf{x}$ vector and the superscription t stands for the transpose of a column vector to a row vector, which can be expressed in a row:

$$\mathbf{x} = (x_1, x_2, \ldots, x_i, \ldots, x_n)^t \qquad (2.1)$$

A matrix is a two-dimensional expression of data, which has a number of rows and a number of columns. For instance, the following matrix has three rows and four columns, or its row dimension is three and its column dimension is four. The dimension of the matrix is thus three by four.

$$\mathbf{A} = \begin{pmatrix} 2 & 3 & 4 & 5 \\ 4 & 6 & 8 & 9 \\ 6 & 9 & 10 & 11 \end{pmatrix} \qquad (2.2)$$

2.2 Operations on vectors and matrices

A vector can be used to take part in different calculations, such as scaling and product. The length (or dimension) of a vector is defined as the number of entries contained in the vector. Scaling a vector means enlarging or reducing all the entries within the vector uniformly. The following notation shows that all the entries of the $\mathbf{x}$ vector have been scaled by a scalar number C:

$$\mathbf{y} = C\mathbf{x} \qquad (2.3)$$

Afterward, each entry of $\mathbf{x}$ is scaled by C, leading to a new vector $\mathbf{y}$ and each entry is changed by the following definition:

$$y_i = Cx_i \qquad (2.4)$$

where x_i and y_i are used to denote the ith entries of the vectors $\mathbf{x}$ and $\mathbf{y}$, respectively. The size or the magnitude of a vector is defined as the norm of a vector and its notation is $\mathcal{L}_\omega$, where the value of ω can be 1 or 2 or ∞ with different outcomes. The $\mathcal{L}_1$-norm of the $\mathbf{x}$ vector is defined as follows, where the norm is the sum of the absolute values of all entries of the $\mathbf{x}$ vector:

$$\mathcal{L}_1(\mathbf{x}) = |\mathbf{x}| = \sum_{i=1}^{n} |x_i| \qquad (2.5)$$

Figure 2.1(a) shows an example of the $\mathcal{L}_1$-norm calculation. The $\mathcal{L}_1$-norm of the vector A is $3 = 1+2$ and the $\mathcal{L}_1$-norm of the vector B

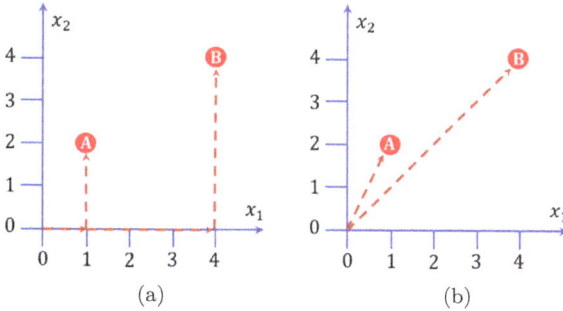

Figure 2.1. An illustration of vector norms. A and B are two vectors. Dashed arrows represent the norm calculation directions. (a) The $\mathcal{L}_1$-norm calculation. (b) The $\mathcal{L}_2$-norm calculation.

is $8 = 4+4$. The $\mathcal{L}_2$-norm of the $\mathbf{x}$ vector is defined as follows, where the norm is the square root of the squared sum of all entries of the vector:

$$\mathcal{L}_2(\mathbf{x}) = \|\mathbf{x}\| = \sqrt{\sum_{i=1}^{n} x_i^2} \qquad (2.6)$$

Note that the $\mathcal{L}_2$-norm of the $\mathbf{x}$ vector is commonly referred to as the Euclidean distance between the vector and the origin of a space. Figure 2.1(b) shows an example of the $\mathcal{L}_2$-norm calculation. The $\mathcal{L}_2$-norm of the vector A is $\sqrt{1+2^2} = \sqrt{5}$ and the $\mathcal{L}_2$-norm of the vector B is $\sqrt{4^2 + 4^2} = 4\sqrt{2}$.

The $\mathcal{L}_\infty$-norm of the $\mathbf{x}$ vector is defined as follows, where the norm is the maximum absolute value of the vector entries:

$$L_\infty(\mathbf{x}) = \max\{|x_i|\} \qquad (2.7)$$

For instance, $\mathcal{L}_\infty((1, 3, -8)) = 8$. A unit vector is such a vector that its $\mathcal{L}_2$ norm is one, i.e., $\|x\| = 1$. The following calculation can transfer an arbitrary vector to a unit vector:

$$\mathbf{u} = \frac{\mathbf{x}}{\|\mathbf{x}\|} \qquad (2.8)$$

It is not difficult to verify the previous equation as shown in the following:

$$\|\mathbf{u}\| = \frac{\|\mathbf{x}\|}{\|\mathbf{x}\|} = 1 \qquad (2.9)$$

The summation and subtraction between two vectors generate a new vector, but the lengths of two vectors must be identical. The distance between two vectors $\mathbf{x}$ and $\mathbf{y}$ is denoted by the following equation:

$$\|\mathbf{x} - \mathbf{y}\| = \sqrt{\sum_{i=1}^{n}(x_i - y_i)^2} \qquad (2.10)$$

For example, the distance between two vectors, A and B, as shown in Figure 2.2 is $\sqrt{(4-1)^2 + (4-2)^2} = \sqrt{13}$. Note that given two vectors $\mathbf{x}$ and $\mathbf{y}$, $\mathbf{x} - \mathbf{y}$ generates a new vector $\mathbf{z}$ whose origin is vector $\mathbf{y}$. For instance, if we move the origin of the space as shown in Figure 2.2 to the point $(1, 2)$, the point (vector) A becomes the origin of a new space and the point (vector) B changes its location from $(4, 4)$ in the old space to $(3, 2)$ in the new space as shown by the dashed arrows coordinated by $z_1 \sim z_2$ in Figure 2.2.

The more frequently used vector operations in machine learning include vector products. They are the dot product operation and the cross-product operation between two vectors. The dot product between two vectors $\mathbf{x}$ and $\mathbf{y}$ is denoted by the following definition, where C is the outcome of the dot product between two vectors $\mathbf{x}$ and $\mathbf{y}$. The result of this dot product is a scalar number:

$$\mathbf{x} \cdot \mathbf{y} = \mathbf{x}^t\mathbf{y} = \mathbf{y} \cdot \mathbf{x} = \mathbf{y}^t\mathbf{x} = C \qquad (2.11)$$

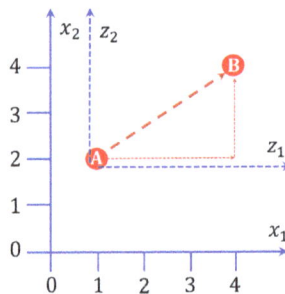

Figure 2.2. An illustration of the distance between two vectors. The original space shown by solid arrows is coordinated by $x_1 \sim x_2$ and the new space shown by the dashed arrows is coordinated by $z_1 \sim z_2$.

For instance, if $\mathbf{x} = (1, 2, 3)^t$ and $\mathbf{y} = (2, 3, 4)^t$, $\mathbf{x} \cdot \mathbf{y} = 20$. This is equivalent to

$$x_1 y_1 + x_2 y_2 + x_3 y_3 \tag{2.12}$$

The dot product operation is also referred to as the projection of one vector on the other. This is based on the trigonometric relationship shown in the following, where α is the angle between two vectors named as $\mathbf{x}$ and $\mathbf{y}$ as shown in Figure 2.3:

$$\mathbf{x} \cdot \mathbf{y} = \|\mathbf{x}\| \|\mathbf{y}\| \cos(\alpha) \tag{2.13}$$

Note that the angle between two vectors satisfies the following condition:

$$\cos(\alpha) = \frac{\mathbf{x} \cdot \mathbf{y}}{\|\mathbf{x}\| \|\mathbf{y}\|} \tag{2.14}$$

The dot product between two vectors can be visualized in Figure 2.3, where the product result is

$$\mathbf{x} \cdot \mathbf{y} = (\|\mathbf{y}\| \cos(\alpha)) \times \|\mathbf{x}\| \tag{2.15}$$

Note that $\|\mathbf{y}\| \cos(\alpha)$ is referred to as the projected norm with respect to vector $\mathbf{x}$. Figure 2.3(a) shows the shape of the original vectors $\mathbf{x}$ and $\mathbf{y}$ in a two-dimensional space. Figure 2.3(b) shows the conversion of the space so that the $\mathbf{x}$ vector is aligned horizontally. Figure 2.3(c) shows how the $\mathbf{y}$ vector is projected onto the $\mathbf{x}$ vector. Finally, the dot product between the $\mathbf{x}$ vector and the $\mathbf{y}$ vector is

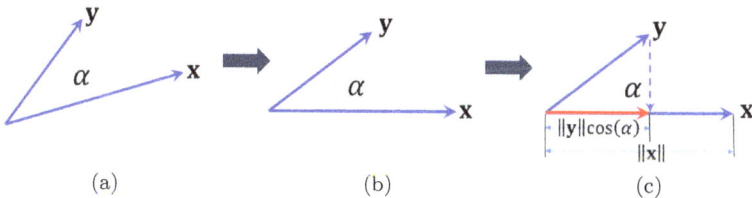

(a)   (b)   (c)

Figure 2.3. An illustration of the dot product between two vectors $\mathbf{x}$ and $\mathbf{y}$. (a) The original positions of two vectors in a two-dimensional space. (b) The conversion of the space so that the $\mathbf{x}$ vector is aligned horizontally. (c) The $\mathbf{y}$ vector is projected on the $\mathbf{x}$ vector through the recognition of the angle between two vectors, i.e., α. The projected norm for $\mathbf{y}$ with respect to $\mathbf{x}$ is $\|\mathbf{y}\| \cos(\alpha)$.

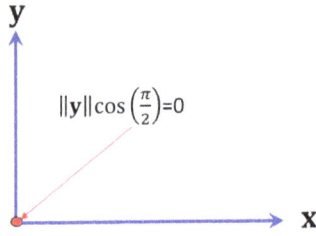

Figure 2.4. An illustration of the dot product between two orthogonal vectors **x** and **y**.

calculated using Eq. (2.15), i.e., the multiplication between the norm of **x** and the projected norm of **y**. There is not doubt that the dot product between two orthogonal vectors (one vector is perpendicular to the other) is zero as shown in Figure 2.4.

The cross-product operation between two vectors is denoted by the following definition and its outcome is a matrix:

$$\mathbf{A} = \mathbf{x}\mathbf{y}^t \tag{2.16}$$

In the abovementioned notation, **A** is a matrix and a brief introduction of matrices will follow in this chapter. If $\mathbf{x} = (1,2,3)^t$ and $\mathbf{y} = (2,3,4)^t$,

$$\mathbf{x}\mathbf{y}^t = \begin{pmatrix} 1 \\ 2 \\ 3 \end{pmatrix} \begin{pmatrix} 2 & 3 & 4 \end{pmatrix} = \begin{pmatrix} 2 & 3 & 4 \\ 4 & 6 & 8 \\ 6 & 9 & 12 \end{pmatrix} \tag{2.17}$$

Note that the position of vectors used in a cross-product operation is important. This is because their results are different. For instance,

$$\mathbf{x}\mathbf{y}^t = (\mathbf{y}\mathbf{x}^t)^t \neq \mathbf{y}\mathbf{x}^t = (\mathbf{x}\mathbf{y}^t)^t \tag{2.18}$$

An important issue regarding cross-product operation is that two vectors do not have to have the same length. For instance, the cross-product between vectors $\mathbf{x} = (1,2,3)^t$ and $\mathbf{y} = (4,5)^t$ leads to

$$\mathbf{x}\mathbf{y}^t = \begin{pmatrix} 4 & 5 \\ 8 & 10 \\ 12 & 15 \end{pmatrix} \tag{2.19}$$

There are some operations between vectors and matrices as well as operations between matrices. Scaling a matrix or multiplying a matrix by a scalar number has the same impact as that in vectors, i.e., all entries of a matrix will be scaled by a scalar number. When the $\mathbf{A}$ matrix is multiplied by the $\mathbf{x}$ vector, the column dimension of the $\mathbf{A}$ matrix must be the same as the length of the $\mathbf{x}$ vector. The multiplication between them generates a new vector $\mathbf{y}$, whose length equals the row dimension of the $\mathbf{A}$ matrix:

$$\mathbf{y} = \mathbf{A}\mathbf{x} \qquad (2.20)$$

The multiplication between a matrix and a vector can be considered as a collection of dot products or a process of projecting a number of data points onto a projection direction. Therefore, the rows of the $\mathbf{A}$ matrix are data points and the $\mathbf{x}$ vector can be treated as the projection direction. This operation has wide applications in linear regression models. Figure 2.5 shows such an example with four data points, which are represented by four vectors in the $\mathbf{A}$ matrix

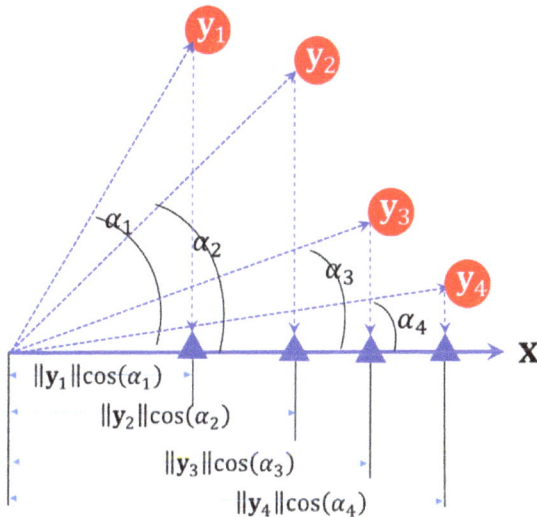

Figure 2.5. An illustration of the multiplication between a matrix and a vector. The filled circles with labels represent the original positions of four data points. The triangles represent the projected positions of these four data points onto the projection direction, i.e., the $\mathbf{x}$ vector.

shown in the following:

$$\mathbf{A} = \begin{pmatrix} \mathbf{y}_1 \\ \mathbf{y}_2 \\ \mathbf{y}_3 \\ \mathbf{y}_4 \end{pmatrix} \tag{2.21}$$

These four data points are projected to the direction of the $\mathbf{x}$ vector. In the real-world applications, the $\mathbf{x}$ vector is replaced by a weight vector $\mathbf{w}$ to construct linear regression models. The summation and subtraction of two matrices require two matrices to have the same dimension. They generate a new matrix with the same dimension. Multiplication between two matrices still generates a matrix. To ensure valid multiplication between two matrices as shown in the following, the column dimension of the $\mathbf{A}$ matrix must be the same as the row dimension of the $\mathbf{B}$ matrix:

$$\mathbf{X} = \mathbf{AB} \tag{2.22}$$

A squared matrix means that its row dimension equals its column dimension. A diagonal matrix, which is a squared matrix as well, is defined as shown in the following, where all off-diagonal entries are zero:

$$\mathcal{D} = \begin{pmatrix} d_{11} & 0 & 0 & \cdots & 0 \\ 0 & d_{22} & 0 & \cdots & 0 \\ 0 & 0 & d_{33} & \cdots & 0 \\ \vdots & \vdots & \vdots & \ddots & \vdots \\ 0 & 0 & 0 & \cdots & d_{nn} \end{pmatrix} \tag{2.23}$$

An orthogonal matrix $\mathbf{O}$ satisfies the following condition:

$$\mathbf{O}^t \mathbf{O} = \mathcal{D} \tag{2.24}$$

An identity matrix, is another squared matrix, and is defined as shown in the following, where the diagonal entries are ones and off-diagonal entries are zeros:

$$\mathbf{I} = \begin{pmatrix} 1 & 0 & 0 & \cdots & 0 \\ 0 & 1 & 0 & \cdots & 0 \\ 0 & 0 & 1 & \cdots & 0 \\ \vdots & \vdots & \vdots & \ddots & \vdots \\ 0 & 0 & 0 & \cdots & 1 \end{pmatrix} \tag{2.25}$$

Sometimes, a subscript is attached to an identity matrix label to indicate the dimension of the matrix. For instance, $\mathbf{I}_3$ represents an identity matrix of dimension three and $\mathbf{I}_n$ represents an identity matrix of dimension n.

The inverse of a matrix $\mathbf{A}$ is denoted by $\mathbf{A}^{-1}$. In machine learning, the inverse of a matrix is widely used. For instance, when analyzing a multivariate Gaussian distribution, the calculation of the inverse of a covariance matrix is required. In the following equation, $\mathbf{u}$ is a mean vector, Σ is the covariance matrix, which is always a squared matrix, and $|\Sigma|$ is the determinant of Σ:

$$\frac{1}{(2\pi)^{d/2}\sqrt{|\Sigma|}} \exp\left(-\frac{1}{2}(\mathbf{x}-\mathbf{u})^t \Sigma^{-1}(\mathbf{x}-\mathbf{u}) \right) \tag{2.26}$$

The multiplication of the $\mathbf{A}$ matrix and its inverse is always an identity matrix:

$$\mathbf{I} = \mathbf{A}^{-1}\mathbf{A} \tag{2.27}$$

One operation on a squared matrix $\mathbf{A}$ converts the matrix to a scalar, which is called matrix determinant $|\mathbf{A}|$. The scalar or the determinant of a matrix has a geometrical meaning, i.e., the area spanned by the specification of a matrix. The inverse of a diagonal matrix is straightforward, i.e.,

$$\mathcal{D}^{-1} = (d_{11}^{-1} \quad d_{22}^{-1} \quad d_{33}^{-1} \quad \cdots \quad d_{nn}^{-1}) \tag{2.28}$$

In addition to the abovementioned vector/matrix operations, some statistical operations applied to vectors/matrices are also useful for the introduction of bio-kernel machines. The mean of a vector $\mathbf{x}$ is defined as follows, where $\mathbf{i}_n = (1 \quad 1 \quad \ldots \quad 1)$ is a unity vector of dimension or length n:

$$\mu = \bar{\mathbf{x}} = \frac{1}{n}\sum_{i=1}^{n} x_i = \frac{1}{n}\mathbf{x} \cdot \mathbf{i}_n \tag{2.29}$$

The variance of a vector $\mathbf{x}$ is defined as follows:

$$\sigma^2 = \frac{1}{n-1}\sum_{i=1}^{n}(x_i - \bar{\mathbf{x}})^2 = \frac{1}{n-1}(\mathbf{x} - \bar{\mathbf{x}})^t(\mathbf{x} - \bar{\mathbf{x}}) \tag{2.30}$$

These statistics applied to vectors can also be applied to matrices. For a matrix $\mathbf{X}$, whose rows represent records and columns represent variables, a covariance matrix can be used to measure the relationship between variables. To do this, the first statistic is the mean vector. The collection of mean vectors for a matrix is defined as follows, where m stands the column dimension of the $\mathbf{X}$ matrix:

$$(\bar{\mathbf{x}}_1, \bar{\mathbf{x}}_2, \ldots, \bar{\mathbf{x}}_m) \tag{2.31}$$

The covariance matrix for the $\mathbf{X}$ matrix is defined as follows, where E stands for the expectation:

$$\mathrm{var}(\mathbf{X}) = \mathrm{E}[(\mathbf{X} - \bar{\mathbf{x}}_i)(\mathbf{X} - \bar{\mathbf{x}}_i)^t] \tag{2.32}$$

For more content on linear algebra, one can refer to Meyer (2000).

Chapter 3

Traditional Kernel Machines

The kernel approach is a class of cutting-edge, advanced machine learning algorithms used for designing and developing various kernel machines. It plays an important role in machine learning due to its powerful function of dealing with many real-world complex problems. The most basic principle of the kernel approach is its capability of mapping a complex feature space to a less complex kernel space for modeling and analysis. Such a space can be linearly separable or closer to linearly separable for a classification/discriminant analysis problem. This chapter starts up a brief introduction of the kernel trick, which is the core of the kernel approach. Since the kernel trick is based on vector operations and kernel machines can also be referred to as vector machines, this chapter will briefly introduce several classical machine learning algorithms, which are based on vector projection in a vector space, and their kernelized versions. Through these introductions, the basic principle of the kernel approach and various kernel machines designed for different real-world applications will be briefly presented.

3.1 Relationship between objects in terms of vector representation

Examining the data distribution pattern of a multi-dimensional data space, by evaluating the mutual relationship between objects, which are represented by vectors in a vector space, based on the pairwise similarity (or distance) measurements between the objects, has been a classical approach. It has laid the cornerstone for machine learning. The relationship between two objects with different physical or biological backgrounds is thus the geometrical relationship between

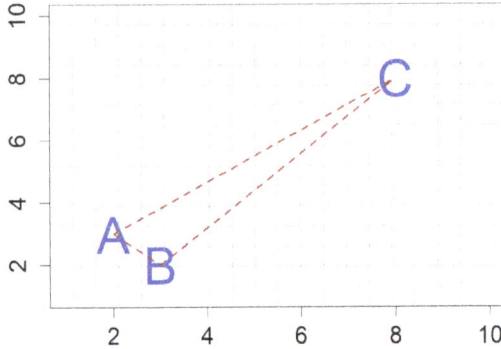

Figure 3.1. An illustration of the relationship between objects (data points and hence vectors) in terms of the distances calculated. The three objects labeled A, B, and C have different mutual distances and hence relationships.

two representative vectors in a vector space. Suppose there are many kinds of objects represented by vectors in a vector space. The mutual relationship between them can be explored or learned through a learning process so as to discover and research the pattern of how data are distributed in a vector space. The learned pattern can thus be used to uncover the unknown truth and for further inference or prediction.

Figure 3.1 shows the mutual relationship between three objects represented by three vectors in a two-dimensional vector space. A and B have a closer relationship. Therefore, they have a shorter distance (or a greater similarity). However, A and C or B and C are more distantly related. Therefore, they have a greater distance (or a smaller similarity). Table 3.1 shows the calculated distances between these three vectors, where the numerical data prove this observation. Therefore, using vectors to quantify the quantitative relationship between multi-dimensional objects is a reliable approach for pattern discovery and research.

In addition to the use of distance measurement or similarity measurement, there is a more useful approach named vector projection using the dot product operation to measure the relationship between objects (vectors) in a vector space. Figure 3.2(a) shows the projection of three vectors onto the projection direction denoted by

Table 3.1. An example to show the distances between the three vectors shown in Figure 3.1.

Points			Distance	
A	2	3	A-B	1.41
B	3	2	A-C	7.81
C	8	8	B-C	7.81

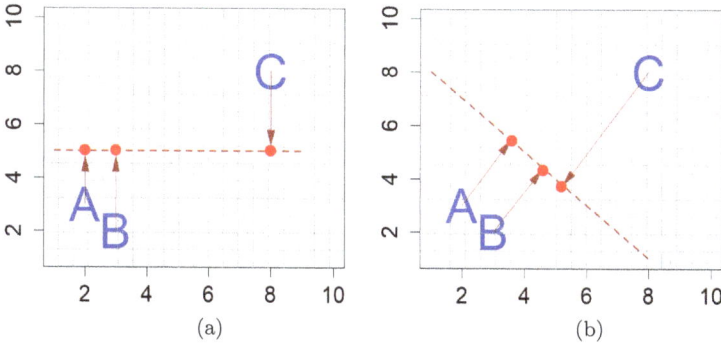

Figure 3.2. An illustration of the relationship measurements between vectors through vector projection. The three objects are labeled A, B, and C. The dotted line represents the projection direction and the arrows represent the projections. (a) A proper projection direction. (b) An improper projection direction.

the dotted line, where it can be seen that the distance between A and B is smaller than the distance between A and C as well as the distance between B and C. This is the basic concept employed in the Fisher discriminant analysis algorithm, which will be discussed in the next section along with other machine learning algorithms including kernel machines. Importantly, the condition to properly and accurately measure the mutual relationship between vectors through vector projection is the correct design or estimation of the projection direction. For instance, Figure 3.2(b) shows an improper projection for the same data, where the distribution of the projected points cannot reflect the true relationship between these three data points.

Another vector projection approach is mutual vector projection using the dot product operation. Table 3.2 shows an example.

Table 3.2. Mutual vector projects between the three vectors shown in Figure 3.1.

	A	B	C
A	13	12	40
B	12	13	40
C	40	40	128

The result of using mutual vector projection between the three data points shown in Table 3.1 is shown in Table 3.2. It can be seen that the result is consistent with the previous discussion

3.2   Data space, feature space, kernel space, and the kernel trick

A space in which a raw dataset has been collected is referred to as a data space or an input space. For instance, the primary structure of a protein is defined as a sequential chain of amino acids, and a collection of sequences is treated as a sequence input space. An image is composed of an array of pixel densities, and a collection of images is treated as an image input space. An input space may not be the main target for a machine learning algorithm to analyze because the data may have no generalizable variables or numerical variables. For instance, two protein sequences in a sequence space may have different lengths and their compositions are non-numerical amino acids. Most machine learning algorithms can only analyze data with generalizable or numerical variables. An input space can be transferred to a new space referred to as a feature space using a mapping process, which is usually a feature extraction process. The design of a feature extraction approach varies with applications. Importantly, each feature in a feature space is observable, measurable, and can be described by numerical values, which can directly be used by a machine learning algorithm. Importantly, each feature in a feature space, once it is created, is generalizable across all data points. For instance, the frequency of a fixed length word will be a generalizable feature no matter whether a molecular sequence is long or short. In the following are three illustrative DNA sequences, which

Table 3.3. The frequency of the 2-mer words for the three illustrative DNA sequences shown in the next page. Only half of the 16-frequency 2-mers are shown due to the table size limit.

	AA	AC	AG	AT	CA	CC	CG	CT
Sequence 1	1	3	4	3	6	5	2	4
Sequence 2	3	2	2	2	1	3	3	4
Sequence 3	4	2	2	3	0	1	0	3

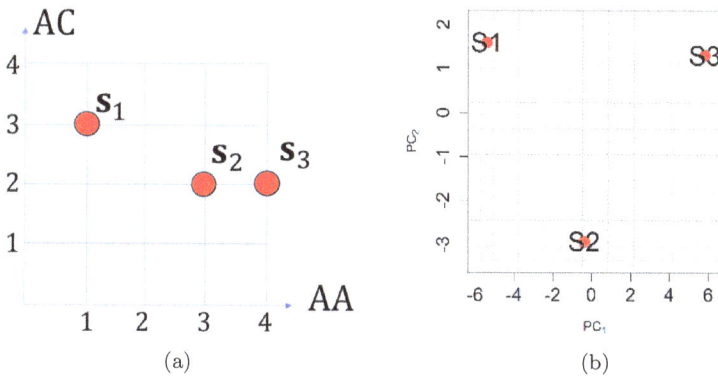

Figure 3.3. The illustration of the feature space for the 2-mer frequencies of the three sequences shown in the next page. (a) The feature space using the first two 2-mer frequencies for the three sequences. (b) The principal component analysis map of the feature space using all 2-mer frequencies for three sequences.

have different lengths. However, the frequency of the k-mer words can be easily estimated for them. Table 3.3 shows the frequency of 2-mer words for these three sequences. Although the three sequences presented in the input space are non-numeric, their 2-mer frequency data presented in the corresponding feature space are numerical and generalizable. Such a numerical space can be geometrically expressed for visualization of how data are distributed and what the qualitative pattern is. For instance, Figure 3.3(a) shows the feature space using the first two 2-mer frequency features for the sequences. Figure 3.3(b) shows the two-dimensional principal component analysis (which will be discussed in Chapter 9) map of the feature space generated based on all 16 2-mer features of the three sequences, where the mutual relationship between the sequences can be well understood.

TAGGGTGCATTTCAGCTCCCATCGCCCCTCACGATACTGACAGCAAGGCT
TAAGCTAGGAACCCGCTTCGACGGCTGATGGTCCTCAATT
ATAACTAGTAGATATCTAAACCTTTAGGAA

A feature space transformed from an input space through a feature extraction procedure may have a complex pattern, making the pattern discovery process difficult. Another mapping process, which is referred to as kernel mapping, can be used to transform such a feature space further to a new space, referred to as a kernel space. It is important to ensure that the variables or coordinates in a kernel space may not be observable or measurable. Figure 3.4 shows an example where a nonlinearly separable feature space is mapped to a kernel space which is linearly separable. In this example, the radial basis function as one popular kernel function, which will be discussed later in this chapter, is used to quantify the relationship between all data points and two randomly selected data points. The quantitative relationship is thus mapped to a kernel space using the radial basis function. Figure 3.4(a) shows the input space, where x_1 and x_2 are two variables in a feature space. It is obvious that

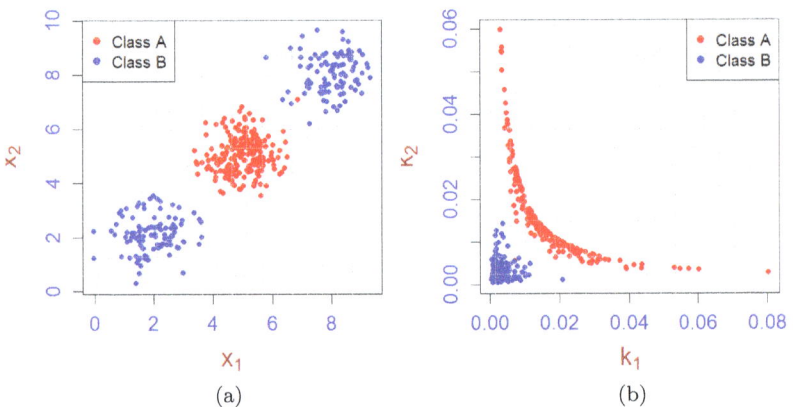

Figure 3.4. The transformation from a feature space which is nonlinearly separable to a kernel space which is linearly separable using the radial basis function. (a) The feature space. (b) The kernel space.

this space is nonlinearly separable. Suppose two vectors (points) are selected from this dataset to serve as two kernels, i.e., acting as the centers of the radial basis function. The two kernels thus represent two coordinates of the kernel space. All data points are transformed to a new space through the relationship (similarity) measurements between them and the two kernels using the radial basis functions. This means that each vector in this feature space will have two similarity measurements regarding the kernels and these two similarity measurements represent a new vector or a new data point in this new space. Figure 3.4(b) shows such a mapping space, where k_1 and k_2 are two new coordinates (kernels) which may not be observable and measurable when we collect the data as shown in Figure 3.4(a). It can be seen that the new space is linearly separable. This simple example shows the advantage of using a number of kernels to map an input space which may be nonlinearly separable to a kernel space which is linearly separable or closer to linearly separable.

Having had a view of how the data distributional characteristic of a feature space can be changed when it is mapped to a kernel space, we now review some properties of the kernel approach, especially the kernel trick, which is the most important concept in kernel machines.

We first review how to generate a kernel space using a kernel function. Suppose a d-dimensional feature space of N data points is denoted by $\mathbf{X} \in \mathcal{R}^{N \times d}$ and the H-dimensional kernel space is denoted by $\mathbf{K} \in \mathcal{R}^{N \times H}$ if H data points serve as the hypothetical kernel vectors or hypothetical kernels. The use of the word "hypothetical" in this book is to emphasize that not all the initially designed kernels will be informative and selected in a modeling process. The dimensionality of the kernel space is thus determined by the number of the hypothetical kernels employed for a mapping from a feature space to a kernel space. To implement such a mapping, a mapping function, which is called the kernel function, is used in most kernel machines. The kernel function $\mathcal{K}$ can be denoted as follows:

$$\mathcal{K} : \mathbf{X} \Rightarrow \mathbf{K} \tag{3.1}$$

The basis of most kernel-based machine learning algorithms is called the kernel trick. We now show some simple cases of how the kernel trick works. Suppose a nonlinear model with a single independent feature variable x is shown as follows:

$$y = \sum_{i=0}^{4} w_i x^i = w_0 x^0 + w_1 x + w_2 x^2 + w_3 x^3 + w_4 x^4 \qquad (3.2)$$

If we need to use a linear method to analyze this model, what we need is to generate a new model using the following format:

$$y = \sum_{i=0}^{4} w_i \phi_i = w_0 \phi_0 + w_1 \phi_1 + w_2 \phi_2 + w_3 \phi_3 + w_4 \phi_4 \qquad (3.3)$$

where there are four variables, i.e., $\phi_i = x^i$ ($\forall i \in \{1, 2, 3, 4\}$) and $\phi_0 = 1$. To establish a linear regression model for Eq. (3.3), a feature matrix is designed as shown in the following:

$$\mathbf{\Phi} = \begin{pmatrix} 1 & \phi_{11} & \phi_{12} & \phi_{13} & \phi_{14} \\ 1 & \phi_{21} & \phi_{22} & \phi_{23} & \phi_{24} \\ 1 & \phi_{31} & \phi_{32} & \phi_{33} & \phi_{34} \\ \vdots & \vdots & \vdots & \vdots & \vdots \\ 1 & \phi_{N1} & \phi_{N2} & \phi_{N3} & \phi_{N4} \end{pmatrix} \qquad (3.4)$$

A model for this problem is thus shown as follows, where $\mathbf{w} = (w_0, w_1, w_2, w_3, w_4)$:

$$\mathbf{y} = \mathbf{\Phi}\mathbf{w} \qquad (3.5)$$

When there are many independent variables and higher polynomial terms, the size of $\mathbf{\Phi}$ may become intractable in terms of the computing memory and speed. The regularization approach is one way to analyze such a model for the optimized model parameters $\mathbf{w}$. The error function with a regularization term is shown as follows, where λ is the Lagrange coefficient:

$$e = \frac{1}{2}(\mathbf{y} - \mathbf{\Phi}\mathbf{w})^t (\mathbf{y} - \mathbf{\Phi}\mathbf{w}) + \frac{1}{2}\lambda \mathbf{w}^t \mathbf{w} \qquad (3.6)$$

The solution to such a model is found through minimizing the above error using the so-called maximum likelihood approach. The solution is shown as follows, where $(\mathbf{\Phi}^t\mathbf{\Phi}+\lambda\mathbf{I})^{-1}$ is called the pseudo-inverse:

$$\mathbf{w} = (\mathbf{\Phi}^t\mathbf{\Phi} + \lambda\mathbf{I}_d)^{-1}\mathbf{\Phi}^t\mathbf{y} \tag{3.7}$$

Note that $\mathbf{\Phi}^t\mathbf{\Phi}$ is the variable-wise inner product conducted for all data points. The operation results in a $d \times d$ squared matrix because

$$(\mathbf{\Phi}^t\mathbf{\Phi} + \lambda\mathbf{I}_d)\mathbf{\Phi}^t = \mathbf{\Phi}^t\mathbf{\Phi}\mathbf{\Phi}^t + \mathbf{\Phi}^t\lambda = \mathbf{\Phi}^t(\mathbf{\Phi}\mathbf{\Phi}^t + \lambda\mathbf{I}_N) \tag{3.8}$$

Left-multiplying $(\mathbf{\Phi}^t\mathbf{\Phi} + \lambda\mathbf{I}_d)^{-1}$ on both sides of Eq. (3.8) leads to the following equation:

$$\mathbf{\Phi}^t = (\mathbf{\Phi}^t\mathbf{\Phi} + \lambda\mathbf{I}_d)^{-1}\mathbf{\Phi}^t(\mathbf{\Phi}\mathbf{\Phi}^t + \lambda\mathbf{I}_N) \tag{3.9}$$

Right-multiplying $(\mathbf{\Phi}\mathbf{\Phi}^t+\lambda\mathbf{I}_N)^{-1}$ again on both sides of Eq. (3.9) results in the following equation:

$$\mathbf{\Phi}^t(\mathbf{\Phi}\mathbf{\Phi}^t + \lambda\mathbf{I}_N)^{-1} = (\mathbf{\Phi}^t\mathbf{\Phi} + \lambda\mathbf{I}_d)^{-1}\mathbf{\Phi}^t \tag{3.10}$$

Equation (3.7) can thus be revised as follows:

$$\mathbf{w} = \mathbf{\Phi}^t(\mathbf{\Phi}\mathbf{\Phi}^t + \lambda\mathbf{I}_N)^{-1}\mathbf{y} \tag{3.11}$$

In Eq. (3.11), $\mathbf{\Phi}\mathbf{\Phi}^t \in \mathcal{R}^{N \times N}$ is a point-wise inner product leading to an $N \times N$ squared matrix. Figure 3.5 shows an example with four data points ($\mathbf{x}_1$, $\mathbf{x}_2$, $\mathbf{x}_3$, and $\mathbf{x}_4$) in a two-dimensional space (v_1 and v_2). Table 3.4 shows a variable-wise inner product of that dataset. Table 3.5 shows the point-wise inner product of that dataset. Comparing Tables 3.4 and 3.5, it can be seen that the sizes of the two resulting matrices are different and the coordinates of the two spaces are also different.

With this linear kernel function $\mathcal{K}$, $\mathbf{\Phi}\mathbf{\Phi}^t$ results in a kernel space $\mathbf{K} \in \mathcal{R}^{N \times N}$. Note that $\mathbf{\Phi}\mathbf{\Phi}^t$ generates a kernel space $\mathbf{K}$, but $\mathbf{\Phi}^t\mathbf{\Phi} \in \mathcal{R}^{d \times d}$ does not. This is what the kernel trick stands for. Using this kernel function, a feature space $\mathbf{\Phi} = (\phi_i)_i^N$ is mapped to a kernel space $\mathbf{K} = (\mathcal{K}(\phi_i, \phi_j))_{i,j}^N$ as denoted in the following:

$$\mathcal{K} : \mathbf{\Phi}\mathbf{\Phi}^t \Rightarrow \mathbf{K} \tag{3.12}$$

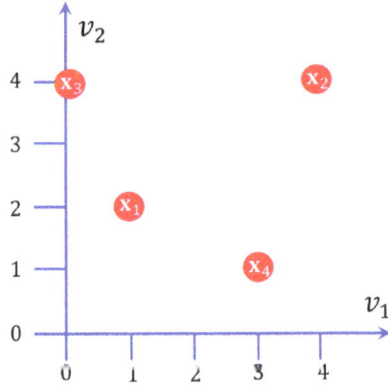

Figure 3.5. An illustration for the matrix inner product.

Table 3.4. The variable-wise inner product result for the data shown in Figure 3.5.

	v_1	v_2
v_1	26	21
v_2	21	37

Table 3.5. The point-wise inner product result for the data shown in Figure 3.5.

	v_1	v_2	v_3	v_4
v_1	5	12	8	5
v_2	12	32	16	16
v_3	8	16	16	4
v_4	5	16	4	10

This mapping leads to the following equation for the estimation of the model parameters $\mathbf{w}$, where $\mathbf{I}_N$ is replaced by $\mathbf{I}$ in the rest of the discussions for simplicity:

$$\hat{\mathbf{w}} = \mathbf{\Phi}^t (\mathbf{K} + \lambda \mathbf{I})^{-1} \mathbf{y} \tag{3.13}$$

When such a model is used for prediction, we have the following equation:

$$\tilde{\mathbf{y}} = \tilde{\mathbf{\Phi}}\mathbf{\Phi}^t(\mathbf{K} + \lambda\mathbf{I})^{-1}\mathbf{y} = \tilde{\mathbf{K}}(\mathbf{K} + \lambda\mathbf{I})^{-1}\mathbf{y} \qquad (3.14)$$

Note that $\tilde{\mathbf{\Phi}} \in \mathcal{R}^{M \times d}$ stands for a feature space of M testing data points, which is different from $\mathbf{\Phi}$. The latter is a set of training vectors and the former is a set of testing vectors. Moreover, $\tilde{\mathbf{y}}$ is also different from $\mathbf{y}$. $\mathbf{y}$ corresponds to a dependent variable and is a set of discrete values for the training data points. If a model is a discriminant problem, $\mathbf{y} \in \{0, 1\}^N$. However, $\tilde{\mathbf{y}} \in \mathcal{R}^M$ corresponds to the predictions for the testing data points and is a set of continuous values. It can be seen that the use of the dot product operation leads to a much simpler model which can be dealt with in a kernel space. In the literature, $\mathbf{\Phi}\mathbf{\Phi}^t$ is referred to as the Gram matrix as shown below, where $\mathcal{K}(\phi_i, \phi_j)$ is implemented by the dot product between data points in this case:

$$\mathcal{K}(\phi_i, \phi_j) = \langle \phi_i, \phi_j \rangle = \phi_i^t \phi_j \qquad (3.15)$$

Suppose vectors in $\mathbf{\Phi}$ are referred to as the training vectors. Equations (3.13) and (3.14) are thus referred to as kernel regression models where the point-wise inner dot product is implemented between new (testing) vectors in $\tilde{\mathbf{\Phi}}$ and the kernel vectors in $\mathbf{\Phi}$ for prediction. Each element in $\tilde{\mathbf{K}}$ is shown as follows:

$$\tilde{k}_{ij} = \tilde{\mathcal{K}}(\tilde{\phi}_i, \phi_j) = \langle \tilde{\phi}_i, \phi_j \rangle = \tilde{\phi}_i^t \phi_j \qquad (3.16)$$

It can be seen that the core of the kernel trick can be used to save time and space. In addition to the use of a point-wise inner dot product or simply a dot product to generate a kernel space, by which a nonlinearly separable space becomes linearly separable or closer to linearly separable, there are also other kernel functions. The polynomial kernel is one of them. Suppose we look at a simple original data space as shown in the following:

$$\phi(x_1, x_2) = (1, x_1, x_2, x_1 x_2, x_1^2, x_2^2) \qquad (3.17)$$

For this data space, we can have two operations. The first is shown in the following:

$$\phi^t(\mathbf{z}_i)\phi(\mathbf{z}_j)$$

$$= 1 + z_{i1}z_{j1} + z_{i2}z_{j2} + z_{i1}z_{i2}z_{j1}z_{j2} + z_{i1}^2z_{j1}^2 + z_{i2}^2z_{j2}^2 \quad (3.18)$$

The second is the polynomial kernel function as shown in the following:

$$(1 + \mathbf{z}_i^t\mathbf{z}_j)^2 = (1 + (z_{i1}, z_{i2})^t(z_{j1}, z_{j2}))^2$$

$$= 1 + 2(z_{i1}, z_{i2})^t(z_{j1}, z_{j2}) + ((z_{i1}, z_{i2})^t(z_{j1}, z_{j2}))^2$$

$$= 1 + 2z_{i1}z_{j1} + 2z_{i2}z_{j2} + z_{i1}z_{i2}z_{j1}z_{j2}$$

$$+ z_{i1}^2z_{j1}^2 + z_{i2}^2z_{j2}^2 \quad (3.19)$$

It can be seen that the two operations are the same except for a slight difference in coefficients. The next commonly used kernel function is the Gaussian kernel function or the radial basis kernel function as shown in the following:

$$\mathcal{K}(\mathbf{z}_i, \mathbf{z}_j) = \exp\left(-\frac{\|\mathbf{x}_i - \mathbf{x}_j\|^2}{2\sigma^2}\right) \quad (3.20)$$

We can use the Taylor series to express this Gaussian kernel function as shown in the following:

$$\exp(x) = \sum_{n=0}^{\infty} \frac{x^n}{n!} = 1 + x + \frac{x^2}{2!} + \frac{x^3}{3!} + \cdots \quad (3.21)$$

Ignoring the high-order terms and assuming $2\sigma^2 = 1$, we have the following approximation to the Gaussian kernel function, which is similar to the outputs of the previous kernel functions:

$$\exp\left(-\frac{\|\mathbf{x}_i - \mathbf{x}_j\|^2}{2\sigma^2}\right) \approx 1 + \|\mathbf{x}_i - \mathbf{x}_j\|^2 \quad (3.22)$$

The condition for a valid kernel function is called Mercer's condition. It requires the following condition to be satisfied. Suppose there is a dataset $(\phi_1, \phi_2, \ldots, \phi_N)$ and two arbitrary numbers $(\alpha_1$ and $\alpha_2)$, then Mercer's condition is defined as follows:

$$\sum_i \sum_j \alpha_1 \alpha_2 \mathcal{K}(\phi_i, \phi_2) \geq 0 \qquad (3.23)$$

3.3 Vector-based learning algorithms and their kernelized versions

A typical representative machine learning algorithm based on vector operations in a multi-dimensional vector space is the K-means algorithm (MacQueen, 1967). The algorithm is used to partition a dataset into a number of clusters. A mean vector (or a central vector) is estimated for each cluster. The algorithm employs a hard membership function to quantify the relationship between data points and a mean vector. Using a hard membership function, the relationship between data points and a mean vector is quantified by a binary number, which is either 0 or 1, representing either no relationship or that they are related, respectively.

Because the centers (mean vectors) are unknown at the beginning of the learning process and are commonly not original data points, the *trial-and-error* learning approach is used by this algorithm and by most other machine learning algorithms. The *trial-and-error* approach is rooted in cognitive science (Campbell, 1960). It seeks the best recognition of the solution (truth) through sequentially repeating similar trials. Its unique property is that each trial in a learning process will make the recognition of the true answer or the true solution better than the previous trials. In more quantitative words, the model will have smaller and smaller errors and hence better and better recognitions in consecutive trials.

The relationship between data points and mean vectors is termed as the membership function or simply the membership in the K-means algorithm. Suppose H hypothetical mean vectors are requested to cluster a dataset using the K-means algorithm. Suppose f_{nh} represents the membership between the nth data point $\mathbf{x}_n \in \mathcal{D} = (\mathbf{x}_1, \mathbf{x}_2, \ldots, \mathbf{x}_N)$ and $\mathbf{u}_h \in \mathcal{H} = (\mathbf{u}_1, \mathbf{u}_2, \ldots, \mathbf{u}_H)$ represents the hth hypothetical mean vector in a hypothesis space $\mathcal{H}$. The membership

f_{nh} is defined as follows:

$$f_{nh} = \begin{cases} 1 & \mathbf{x}_n \text{ has a minimum distance with } \boldsymbol{u}_h \\ 0 & \text{otherwise} \end{cases} \qquad (3.24)$$

The error function of the K-mean algorithm is defined as follows:

$$e = \sum_{h=1}^{H} \sum_{n=1}^{N} f_{nh} \| \mathbf{x}_n - \boldsymbol{u}_h \|^2 \qquad (3.25)$$

Minimizing this error function leads to the adaptive update rule of mean vectors $\boldsymbol{u}_h$ for clusters:

$$\boldsymbol{u}_h^{t+1} = \frac{1}{N_h} \sum_{n=1}^{N} f_{nh}^t \mathbf{x}_n \qquad (3.26)$$

where t is the learning iteration cycle, $\boldsymbol{u}_h^t$ is the mean vector estimated at the cycle t, $\boldsymbol{u}_h^{t+1}$ is the mean vector estimated at the cycle $t+1$, f_{nh}^t is the membership estimated at the cycle t, and $N_h = \sum_{n=1}^{N} f_{nh}^t$. Based on the trial-and-error approach, this updated rule is repeatedly employed until the error is sufficiently small (less than a pre-defined threshold) or the mean vectors become stable (no change) in consecutive learning cycles, such as $\sum_{h=1}^{H} (\boldsymbol{u}_h^{t+1} - \boldsymbol{u}_h^t)^2 < \epsilon$, where ϵ is a predefined threshold for the stability of an estimated model.

Figure 3.6(a) shows a dataset of four clusters with four initial mean vectors. It can be seen that these four initial mean vectors are definitely incorrect. Figure 3.6(b) shows the learning result using the K-means algorithm for this dataset. It can be seen that the four mean vectors have been correctly estimated.

The classical K-means algorithm described earlier employs a hyperplane to separate a data space to generate clusters. In order to deal with more complex data, the K-means algorithm has been kernelized (Dhillon *et al.*, 2004; Dhillon *et al.*, 2007; Tzortzis and Likas, 2009; Liu, 2023). The kernelized K-means algorithm maps a

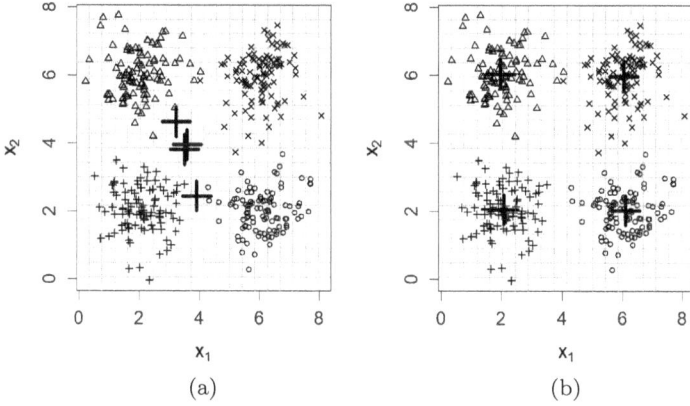

Figure 3.6. An example of applying the K-means algorithm to a dataset of four clusters. The smaller patterns (triangles, crosses, pluses, and open dots) represent the raw data, while the larger pluses represent the mean vectors. (a) Four initial mean vectors. (b) Four mean vectors estimated by the K-means algorithm.

feature space to a high-dimensional kernel space using a kernel function. Suppose an input space is mapped to a new space using $\phi(\mathbf{x}_n)$. The error function is defined as follows:

$$e = \sum_{h=1}^{H} \sum_{n=1}^{N} f_{nh} \|\phi(\mathbf{x}_n) - \boldsymbol{u}_h\|^2 \tag{3.27}$$

The estimated mean vector in this space is defined as follows:

$$\boldsymbol{u}_h = \frac{1}{N_h} \sum_{i=1}^{N} f_{ih} \phi(\mathbf{x}_i) \tag{3.28}$$

Note that $\|\phi(\mathbf{x}_n) - \boldsymbol{u}_h\|^2$ can be expressed as the dot product as shown in the following:

$$\phi(\mathbf{x}_n) \cdot \phi(\mathbf{x}_n) - 2 \cdot \frac{1}{N_h} \sum_{i=1}^{N} f_{ih}^t \phi(\mathbf{x}_i) \cdot \phi(\mathbf{x}_n)$$

$$+ \frac{1}{N_h^2} \sum_{i=1}^{N} (f_{ih}^t)^2 \phi(\mathbf{x}_i) \cdot \phi(\mathbf{x}_i) \tag{3.29}$$

It can be rewritten as follows, where the kernel function $\mathcal{K}$ is employed:

$$\mathcal{K}(\mathbf{x}_n, \mathbf{x}_n) - 2 \cdot \frac{1}{N_h} \sum_{i=1}^{N} f_{ih}^t \mathcal{K}(\mathbf{x}_i, \mathbf{x}_n)$$

$$+ \frac{1}{N_h^2} \sum_{i=1}^{N} (f_{ih}^t)^2 \mathcal{K}(\mathbf{x}_i, \mathbf{x}_i) \qquad (3.30)$$

This discussion shows that the kernelized K-means algorithm can be implemented in a kernel space to analyze more complex data.

Figure 3.7 shows a comparison between a K-means model and a kernelized K-means model constructed for a dataset with two classes of nonlinear data. In Figure 3.7(a), the estimated mean vector found by the K-means model for the outer ring class is obviously inaccurate though the mean vector for the inner class is not too bad. In Figure 3.7(b), two mean vectors have been well estimated by the kernelized K-means model.

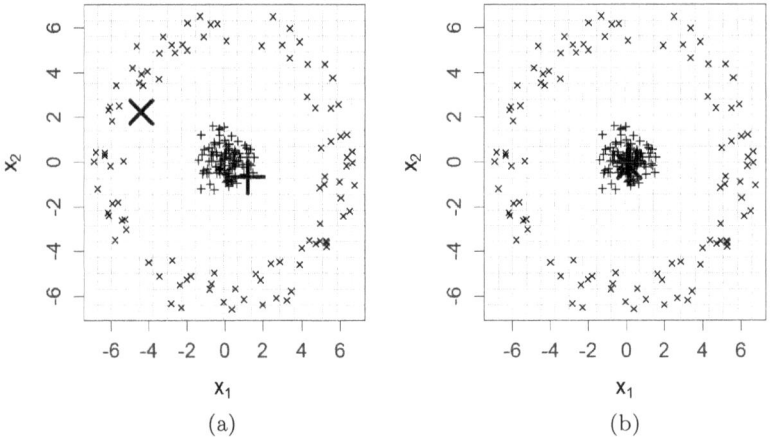

Figure 3.7. An example of applying the kernelized K-means algorithm to a dataset with two classes of data points. The smaller patterns (pluses and crosses) represent the raw data points, while the larger pluses and crosses represent the mean vectors found by the two models. (a) The K-means model outcome. (b) The kernelized K-means model outcome.

Another classical example is the learning vector quantization algorithm (LVQ), which was developed in the 1980s (Burton *et al.*, 1983; Gray, 1984; Soong *et al.*, 1985; Kohonen, 1989; Sato and Yamada, 1995). The LVQ algorithm is also used to partition a space into subspaces (or clusters) and is an adaptive learning algorithm. Similar to the K-means algorithm, the prototype vectors are not necessarily the original data vectors. However, unlike the K-means algorithm, the LVQ algorithm employs a soft membership function. Suppose a dataset with N data points is denoted by $\mathbf{X} = (\mathbf{x}_1, \mathbf{x}_2, \ldots, \mathbf{x}_N) \in \mathcal{R}^{N \times d}$, where d is the dimension of the data space. The error function employed by the LVQ algorithm is shown in the following:

$$e = \sum_{n=1}^{N} f(\mathbf{x}_n) \|\mathbf{x}_n - \rho(\mathbf{x}_n)\|^2 \tag{3.31}$$

where $f(\mathbf{x}_n)$ is a density function at the data point $\mathbf{x}_n$ and $\rho(\mathbf{x}_n) \in \mathcal{H}$ is the prototype vector to represent $\mathbf{x}_n$ in a hypothetical space $\mathcal{H}$. Two optimizing criteria are employed to minimize the objective function defined previously. They are the nearest neighbor constraint (NC) and the centroid constraint (CC). NC requires each subspace to be occupied by at least one data point and CC requires each prototype vector to be a mean vector of all data points which are the nearest neighbor vectors of the prototype vector. To search for a set of prototype vectors, an LVQ model can be constructed using the following weight update rule (Kohonen, 1989; Sato and Yamada, 1995):

$$\mathbf{w}_i^{t+1} = \mathbf{w}_i^t - \alpha^t(\mathbf{x}_n - \mathbf{w}_i^t) \tag{3.32}$$

where $\mathbf{w}_i^t \in \mathcal{R}^d$ is one of the prototypes at the learning cycle t and α^t is a learning parameter called learning rate at the learning cycle t. Figure 3.8 shows an example of applying LVQ. The LVQ algorithm has also been kernelized leading to the kernelized LVQ (kLVQ) algorithm (Satish and Sekhar, 2006).

The self-organizing map (SOM) network (Kohonen, 1989) is a further development of the LVQ algorithm as well as an earlier work regarding self-organizing characteristics (von der Malsburg, 1973).

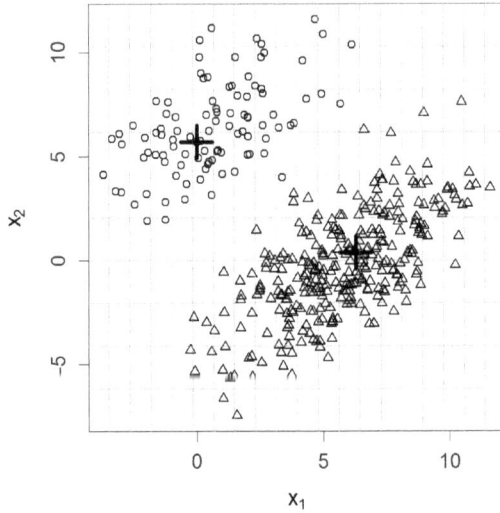

Figure 3.8. An example of applying LVQ to a dataset of two clusters, which are represented by triangles and the circles. The two found prototypes are denoted by the pluses.

A specific feature of SOM is that it employs an array of nodes called output neurons to display or visualize the learning outcome. The SOM algorithm was originally developed for data visualization or topology structure reservation (Kohonen, 1989). The algorithm maps a data space to a two-dimensional space through a competitive learning process. The two-dimensional space is a discrete space in which output neurons or grids are fixed. Figure 3.9 shows an example of an SOM model. It can be seen that such a model does not use the data class label during the learning process because SOM is an unsupervised learning algorithm. Suppose an input vector is denoted by $\mathbf{x} \in \mathcal{R}^d$ and the connection between the output neurons and input neurons is denoted by a matrix $\mathbf{W} \in \mathcal{R}^{M \times d}$, where d is the dimension of a data space and M is the number of output neurons of an SOM model. Because SOM is an unsupervised learning algorithm, its learning mechanism is thus a process of the recognition process of the topological structure hidden in a dataset. SOM adopts an online learning algorithm, i.e., the model parameters (weights shown in Figure 3.9) are updated whenever an input is fed into the model.

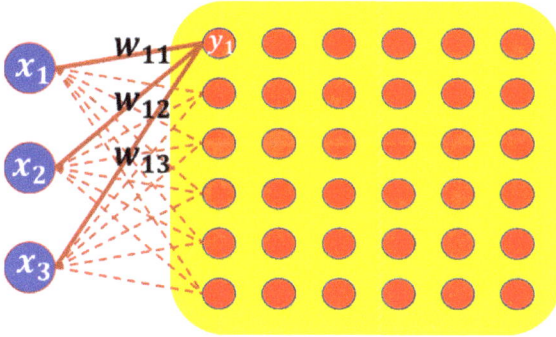

Figure 3.9. An example of an SOM model, which has three input neurons (x_1, x_2, and x_3) and 6×6 output neurons. Each input neuron is associated with one input variable. All the output neurons are fixed in grids as a 6×6 array. Each output neuron is fully connected with all the input neurons quantitatively. The connections between the first output neuron (y_1) and three input neurons are labeled by w_{11}, w_{12}, and w_{13}. Only the connections between the first column of the output neurons and the input neurons are drawn in this plot for simplicity.

For instance, the weights (w_{11}, w_{12}, and w_{13}) connecting the first output neuron (x_1, x_2, and x_3) and the input neurons will be updated depending on their distance.

The construction of an SOM model is implemented by minimizing the distance between model parameters and data vectors. For M output neurons, there are therefore M distances (errors) for each input vector $\mathbf{x}_n$. A winner output neuron or a winner is sought through minimizing the M distances between this input and all weight vectors ($\mathbf{w}_m$). The winner is denoted by π_n and is defined as follows:

$$\pi_n = \underset{m \in [1,M]}{\mathrm{argmin}} \left\{ \frac{1}{2} \|\mathbf{x}_n - \mathbf{w}_m\|^2 \right\} \tag{3.33}$$

This definition implies the following optimization:

$$\mathbf{w}_\pi = \underset{\mathbf{w}_m \in \mathbf{w}}{\min} \left\{ \frac{1}{2} \|\mathbf{x}_n - \mathbf{w}_m\|^2 \right\} \tag{3.34}$$

Based on the detected winner for each input, the weight vectors are updated, and the SOM update rule of the weight vectors is defined

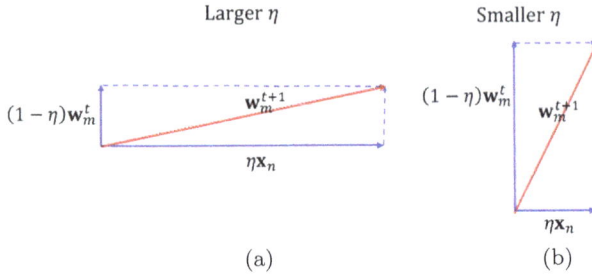

Figure 3.10. An illustration of the SOM learning mechanism. It shows how a weight vector moves toward the direction of an input vector. (a) When the learning rate is larger. (b) When the learning rate is smaller.

as follows, where $0 < \eta < 1$ is referred to as a learning rate:

$$\Delta \mathbf{w}_m = -\eta \frac{\partial \frac{1}{2} \|\mathbf{x}_n - \mathbf{w}_m\|^2}{\partial \mathbf{w}_m} = \eta(\mathbf{x}_n - \mathbf{w}_m) \qquad (3.35)$$

The abovementioned update rule is finalized as follows:

$$\mathbf{w}_m^{t+1} = \mathbf{w}_m^t + \eta(\mathbf{x}_n - \mathbf{w}_m^t) = \eta\mathbf{x}_n + (1 - \eta)\mathbf{w}_m^t \qquad (3.36)$$

Figure 3.10 shows this learning SOM mechanism. The learning rate controls the moving direction of the weight vector. When the learning rate is larger, the major contribution to the weight vector comes from the input vector. This means that there will be a greater move for the weight vector toward the input vector. Figure 3.10(a) shows this scenario. When the learning rate is smaller, the major contribution to the weight vector comes from the weight vector itself. Therefore, the update on the weight vector is small. Figure 3.10(b) shows this scenario. This learning mechanism shows an important feature, i.e., the topological structure of a dataset can be well learned and reserved through such a learning process. This is because the weight vectors always move toward the inputs no matter whether the learning rate is small or large.

The major feature of an SOM is competitive learning in addition to the online learning mechanism. The use of the competitive learning mechanism ensures the robustness of the learning process using SOM. To be more specific, in a learning process using SOM, it can

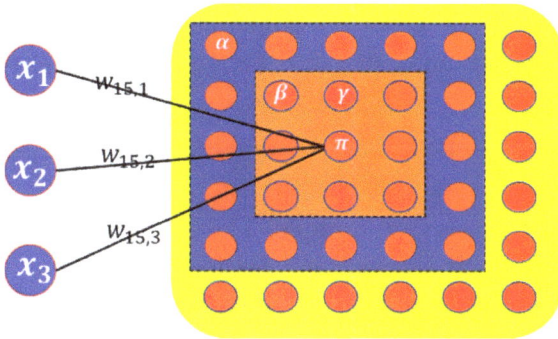

Figure 3.11. An illustration of the competitive learning mechanism based on the neighborhood design in SOM. The winner neuron is denoted by π. This neighborhood structure is composed of two layers. The first layer of the neighborhood is composed of eight neurons which are the closet to the winner neuron. The second layer of the neighborhood is composed of 16 neurons beyond the first layer of the neighborhood, but within the neighborhood structure.

be difficult to avoid oscillation without this competitive learning mechanism. Competitive learning forces the update of the weight vector to be competitive to the winner. A winner is the output neuron whose weight vector has the minimum distance to an input vector. This competitive learning mechanism is also referred to as the winner-take-all strategy. To implement competitive learning, the neighborhood principle is thus designed in the SOM. Its mechanism is shown in Figure 3.11, where there are two layers of the neighborhood. In real data analysis, there might be more than two neighborhood layers. The first layer of the neighborhood is composed of eight neurons which are the closest to the winner (π_n). π_n stands for the winner for the nth input vector. The second layer of the neighborhood is composed of 16 neurons which are beyond the first layer of the neighborhood but within the neighborhood structure for an input vector.

Having introduced the neighborhood concept for competitive learning, the weight update rule of SOM is revised as follows, where $\mu_{\pi m}$ is the neighborhood effect:

$$\mathbf{w}_m^{t+1} = \mathbf{w}_m^t + \eta \mu_{\pi m}(\mathbf{x}_n - \mathbf{w}_m^t) \tag{3.37}$$

The neighborhood effect is defined as follows, where $\mathcal{L}_{\mathcal{H}}(\pi_n, m)$ stands for the Hamming distance (or the $\mathcal{L}_1$ norm) regarding the winner π_n:

$$\mu_{\pi_n m} = e^{-\mathcal{L}_{\mathcal{H}}(\pi_n, m)} \tag{3.38}$$

Take Figure 3.11 as an example. The Hamming distance between the winner (π_n) and the neuron α is four. The Hamming distance between the winner (π_n) and the neuron β is two, while the Hamming distance between the winner (π_n) and the neuron γ is one.

The third important concept of competitive learning employed by SOM is the decaying learning rate and the decaying neighborhood effect. Combining the two decaying rates results in the final weight vector update rule as shown in the following:

$$\mathbf{w}_m^{t+1} = \mathbf{w}_m^t + \eta^t \mu_{\pi m}^t (\mathbf{x}_n - \mathbf{w}_m^t) \tag{3.39}$$

The decaying effect is shown in the following, where T is the maximum learning cycle which is determined by the user, η_0 is the initial learning rate, and μ_0 is the initial neighborhood size:

$$\eta^t = \eta_0 \left(1 - \frac{t}{T}\right) \quad \mu_{\pi m}^t = \mu_0 \left(1 - \frac{t}{T}\right) \tag{3.40}$$

It can be seen that SOM employs the online competitive learning approach, while the K-means algorithm and the LVQ algorithm employ a non-competitive learning approach. The K-means algorithm and the LVQ algorithm aim to partition a space into subspaces, while SOM aims to explore and reserve the underlying topology structure of a dataset. Figure 3.12 shows an example of applying SOM to a two-dimensional data space, where the small dots represent the input data, while the filled diamonds represent the neuron weights. The lines connecting the neurons in the array are used to visualize the topology structure of the data, i.e., how the 25 clusters are interlinked to each other and organized in an array. This example shows that SOM is a powerful tool for data visualization through the operations in a vector space. Kernelized SOM will be discussed in particular in Chapter 9.

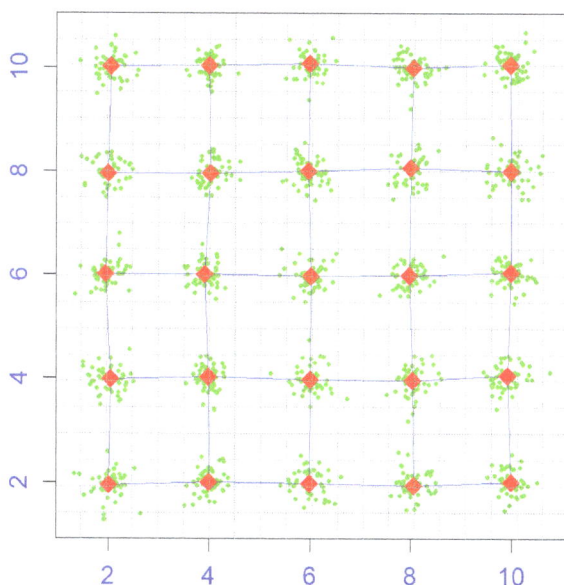

Figure 3.12. An example of applying SOM to reveal the underlying topology structure of a dataset. The small dots represent the input data. The filled diamonds represent the neuron weights and the lines are used to represent the topology structure of the data, which is revealed through an SOM model. The weights can also represent the centers of 25 clusters.

3.4 Fisher discriminant analysis

We now review one of the earliest machine learning algorithms for discriminant analysis, i.e., the Fisher discriminant analysis (FDA) or the linear discriminant analysis algorithm (Fisher, 1936). It lays the foundation for classification/discriminant analysis in the machine learning community and is still very popular in real-world applications, including biology. For instance, it has been employed for the study of the geographical origin of maize (Wang *et al.*, 2020), the study of the differences between the chemotaxonomic and ecophysiological traits of green microalgae fatty acids (Stamenkovic *et al.*, 2020), and the study of the rule of gastrectomy on gastric cancer treatment (Erawijantari *et al.*, 2020).

What it is really interesting regarding the learning approach is how FDA works based on the vector operations in a vector space.

As discussed in the previous chapter, vector projection is based on the dot product operation between two vectors, one of which is treated as the projection direction. The key, and the first step of FDA, is indeed the employment of the vector projection operation, by which all data points represented by vectors are projected onto the direction of one projection vector in a vector space. In this way, a high-dimensional space is mapped to a one-dimensional space. Discriminant analysis is carried out in this one-dimensional space by analyzing the densities of two clusters of the projected data points.

Suppose a dataset is in a d-dimensional space. An object represented by a vector is denoted by $\mathbf{x} = (x_1, x_2, \ldots, x_d)$. In addition, the vector $\mathbf{w} = (w_1, w_2, \ldots, w_d)$ represents the projection direction. By this projection direction, FDA maps an input vector $\mathbf{x}$ to a projected value (a scalar) in a one-dimensional space as shown in the following:

$$\hat{y} = \mathbf{x}^t \mathbf{w} = x_1 w_1 + x_2 w_2 + \cdots + x_d w_d \qquad (3.41)$$

where $\hat{y} \in \mathcal{R}$ is referred to as the value of an input vector $\mathbf{x}$ projected onto the projection direction $\mathbf{w}$. Suppose there are N objects which are represented by N input vectors, $\mathbf{x}_1$, $\mathbf{x}_2$, and $\mathbf{x}_N$. In this circumstance, there will be N dot products to project N vectors onto the projection direction $\mathbf{w}$, resulting in N projected values, i.e.,

$$(\mathbf{x}_1^t \mathbf{w}, \mathbf{x}_2^t \mathbf{w}, \ldots, \mathbf{x}_N^t \mathbf{w}) \rightarrow (\hat{y}_1, \hat{y}_2, \ldots, \hat{y}_N) \qquad (3.42)$$

There is no doubt that these N projected values will spread on a one-dimensional projection space. The distribution of these projected data constitutes a one-dimensional density. FDA is designed for discriminant analysis, i.e., separating two classes of input data points by analyzing the density. Suppose each of these N data points has a unique label y, which is binary, i.e., being either zero or one. What is expected is that $\hat{y}_n$ should approach zero as close as possible if the label of the nth input data point is zero and $\hat{y}_n$ should approach one as close as possible if the label of the nth input data point is one. The label vector of N data points is denoted by $\mathbf{y} = (y_1, y_2, \ldots, y_N)$. The following mapping is thus the objective of a learning process, where one-to-one correspondence should be held, i.e., $\hat{y}_n \mapsto y_n$, $\forall n \in [1, N]$:

$$(\hat{y}_1, \hat{y}_2, \ldots, \hat{y}_N) \mapsto (y_1, y_2, \ldots, y_N) \tag{3.43}$$

It is assumed and also required that the projected values made by FDA constitute two densities. One density corresponds to one class of inputs and the other density corresponds to the other class. Therefore, the main objective of FDA is to search for an optimal projection direction $\hat{\mathbf{w}}$ by which two densities are separated as far as possible.

FDA assumes that each class of the input data follows a multivariate normal distribution. Suppose two mean vectors (centers) of two normally distributed data are denoted by $\boldsymbol{u}_1$ and $\boldsymbol{u}_2$, where $\boldsymbol{u}_1$ corresponds to the data points labelled by class one and $\boldsymbol{u}_2$ corresponds to the data points labelled by class two. The two mapped centers in the projection space, which is a one-dimensional space, can be estimated using the following equation:

$$\tau_1 = \mathbf{w}^t \boldsymbol{u}_1$$
$$\tau_2 = \mathbf{w}^t \boldsymbol{u}_2 \tag{3.44}$$

Note that $\tau_1 \in \mathcal{R}$, $\tau_2 \in \mathcal{R}$ and the distance (variance) between them or the between-class variance is shown as follows:

$$S_B = \tau_2 - \tau_1 = \mathbf{w}^t \boldsymbol{u}_2 - \mathbf{w}^t \boldsymbol{u}_1 = \mathbf{w}^t (\boldsymbol{u}_2 - \boldsymbol{u}_1) \tag{3.45}$$

Moreover, suppose Σ_1 and Σ_2 are the covariance matrices of two classes of inputs in the original multi-dimensional space. The total within-class variance of two classes in the projection space is defined as follows:

$$S_W = \mathbf{w}^t (\Sigma_2 + \Sigma_1) \mathbf{w} \tag{3.46}$$

To measure how two densities in the projection space are well separated, a ratio is used to evaluate the success or goodness of a projection direction, which is defined as follows:

$$R = \frac{S_B}{S_W} = \frac{\mathbf{w}^t (\boldsymbol{u}_2 - \boldsymbol{u}_1)}{\mathbf{w}^t (\Sigma_2 + \Sigma_1) \mathbf{w}} \tag{3.47}$$

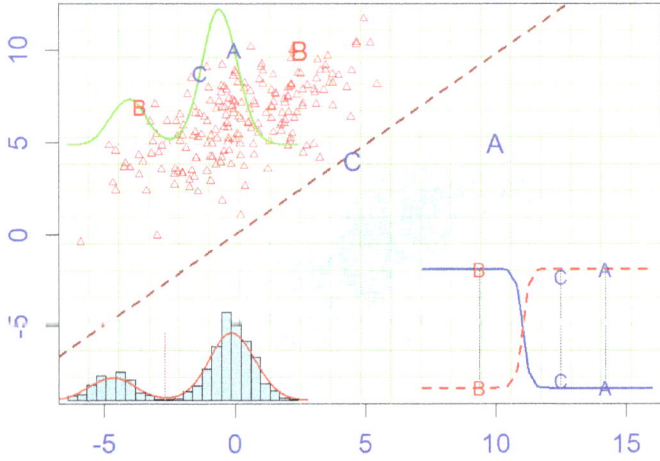

Figure 3.13. An illustration of how FDA works (regenerated from Figure 3.2 of Yang, 2022a). The dots and pluses represent two classes of raw data before being mapped to the projection space. The three spotting points denoted by A, B, and C are used to examine whether they are mapped to the correct positions in the projection space. The projection direction **w** is denoted by a dotted line. The bottom-left inset shows the density of the projected values. The top-left inset shows the positions of spotting points in the density function. The bottom-right inset shows the decision-making curves using the Bayes rule.

FDA searches for the optimal projection direction $\hat{\mathbf{w}}$ by maximizing the abovementioned ratio. Optimization leads to the following equation (Duda *et al.*, 2000):

$$\hat{\mathbf{w}} \propto (\Sigma_2 + \Sigma_1)^{-1}(\boldsymbol{u}_2 - \boldsymbol{u}_1) \qquad (3.48)$$

Figure 3.13 shows an example, where three spotting points are used to show whether an FDA model can accurately project these spotting points to the correct positions in the projection space based on an optimized projection direction $\hat{\mathbf{w}}$. In this plot, the bottom-right inset shows how the Bayes rule is used for decision-making. Suppose two classes of data points before mapping are labeled A and B. The density functions or so-called empirical models in the projection space are denoted by $f_A(\mathbf{x})$ and $f_B(\mathbf{x})$. One thing which is important in Bayesian theory is that some prior knowledge exists before one can collect data to build these empirical models $f_A(\mathbf{x})$ and $f_B(\mathbf{x})$. The prior knowledge, in a simple way, measures the importance or weight of each class in terms of their contribution to a decision-making process using the empirical models. The two prior probabilities of the two classes are normally denoted by π_A and π_B. They satisfy the following condition:

$$\pi_A + \pi_B \equiv 1$$
$$\pi_A, \pi_B \geq 0 \tag{3.49}$$

Based on the empirical models and the prior probabilities, the posterior probabilities (p_A and p_B) for two classes are defined as below:

$$p_A(\mathbf{x}) = \frac{\pi_A f_A(\mathbf{x})}{\pi_A f_A(\mathbf{x}) + \pi_B f_B(\mathbf{x})}$$
$$p_B(\mathbf{x}) = \frac{\pi_B f_B(\mathbf{x})}{\pi_A f_A(\mathbf{x}) + \pi_B f_B(\mathbf{x})} \tag{3.50}$$

Making decisions for a constructed FDA model is based on the Bayes rule shown in the following:

$$\mathbf{x} = \begin{cases} A & p_A(\mathbf{x}) > p_B(\mathbf{x}) \\ B & p_A(\mathbf{x}) < p_B(\mathbf{x}) \end{cases} \tag{3.51}$$

FDA employs one projection vector for a linear problem. Employing more than one projection vector for modeling complex data leads to advanced machine learning approaches (Rosenblatt, 1956; Parzen, 1962; Specht, 1990; Yang and Chen, 1998; Venables and Ripley, 2002; Kowalski and Kusy, 2018; Savchenko, 2020). FDA has also been kernelized for some real-world applications (Mika *et al.*, 1999; Baudat and Anouar, 2000).

3.5 Parzen–Rosenblatt windows

The Parzen–Rosenblatt windows (PRWs) (Rosenblatt, 1956; Parzen, 1962) may be the earliest kernel approach employing more than one vector (kernel) for modeling nonlinear data. Moreover, unlike LVQ and FDA, PRW and other kernel machines discussed later in this chapter employ input data points as the kernels for a learning machine model. PRW is still utilized in real-world applications (Garpebring *et al.*, 2018; Yao *et al.*, 2022).

Suppose there are H independent and identically distributed data points denoted by $(\boldsymbol{u}_1, \boldsymbol{u}_2, \ldots, \boldsymbol{u}_H)$ and they are assumed to be randomly drawn from a dataset. Note that it is assumed that all these H data points have the same contributing role to the unknown density $f(x, w)$. This role is measured by a smoothing parameter w, which is also referred to as the bandwidth in PRW. The unknown density $f(x, w)$ is defined as follows:

$$f(x, w) = \frac{1}{Hw} \sum_{h=1}^{H} P\left(\frac{\|\mathbf{x} - \boldsymbol{u}_h\|}{w}\right) \tag{3.52}$$

where P is the Parzen–Rosenblatt transfer function. The recent implementation of PRW in pattern recognition was in the algorithm referred to as probabilistic neural network (PNN) (Specht, 1990; Yang and Chen, 1998; Yang *et al.*, 2000; Kowalski and Kusy, 2018; Savchenko, 2020). The empirical model or the conditional probability associated with a class k of PNN is defined as follows:

$$P(\mathbf{x}|k) = \sum_{h=1}^{H_k} \omega_h^k f(\mathbf{x}|\boldsymbol{u}_h^k) \tag{3.53}$$

where H_k is the number of hypothetical kernels (referred to as the windows in PRW) employed for the kth class, $\mathbf{x}$ is the inference point of the new vector for the prediction, ω_h^k is the mixing coefficient of the hth kernel in the kth class, $\boldsymbol{u}_h^k$ is the center of the hth Gaussian of the kth class, and $f(\mathbf{x}|\boldsymbol{u}_h^k)$ is the kernel density function. The condition for ω_h^k is shown in the following:

$$\sum_{i=1}^{H_k} \omega_h^k = 1 \quad \& \quad \omega_h^k \geq 0 \tag{3.54}$$

The density function is defined as follows:

$$f(\mathbf{x}|\boldsymbol{u}_h^k) = \frac{1}{(2\pi\sigma_{kh}^2)^{d/2}} \exp\left(-\frac{(\mathbf{x} - \boldsymbol{u}_h^k)^t(\mathbf{x} - \boldsymbol{u}_h^k)}{2\sigma_{kh}^2}\right) \tag{3.55}$$

where σ_{kh}^2 is the variance or the smoothing parameter of the hth kernel in the kth class, d is the data dimension $(\mathbf{x}, \boldsymbol{u}_h^k \in \mathcal{R}^d)$, and k can be either A or B.

One of the procedures to train a PRW or PNN classifier is the maximum likelihood approach. Suppose there is a training dataset with K classes. $(\mathbf{x}_1^k, \mathbf{x}_2^k, \ldots, \mathbf{x}_{N_k}^k)$ is a set of training vectors for the kth class, where N_k is the number of training vectors of the kth class. $(\mathbf{u}_1^k, \mathbf{u}_2^k, \ldots, \mathbf{u}_H^k)$ is a set of center vectors serving as the kernels of kth class. The log-likelihood function of a PRW classifier is thus defined as follows:

$$\log \mathcal{L} = \sum_{k=1}^{K} \sum_{n=1}^{N_k} \log P(\mathbf{x}_n^k | k)$$

$$= \sum_{k=1}^{K} \sum_{n=1}^{N_k} \log \left\{ \sum_{h=1}^{H_k} w_h^k f(\mathbf{x}_n^k | \mathbf{u}_h^k) \right\} \tag{3.56}$$

To estimate the model parameters depending on the condition shown in the previous equation, we can introduce the Lagrange term shown as follows:

$$\log \mathcal{L} = \sum_{k=1}^{K} \sum_{n=1}^{N_k} \log P(\mathbf{x}_n^k | k) + \sum_{k=1}^{K} \lambda_k \sum_{h=1}^{H_k} (w_h^k - 1)^2 \tag{3.57}$$

where λ_k is the Lagrange multiplier of the kth class. Maximizing this objective function leads to the estimation of all model parameters including $\mathbf{u}_h^k$, w_h^k, and σ_{kh}^d (Yang and Chen, 1998). Suppose we deal with a discrimination analysis problem and π_A and π_B are the prior probabilities of two classes. The posterior probability $P(k|\mathbf{x})$ for a data point $\mathbf{x}$ to belong to a class c (A or B) is defined as follows:

$$P(\tau | \mathbf{x}) = \frac{\pi_\tau P(\mathbf{x} | \tau)}{\sum_{k=1}^{K} \pi_k P(\mathbf{x} | k)} \tag{3.58}$$

Note that π_k must satisfy the condition shown in the following:

$$\sum_{k=1}^{K} \pi_k = 1 \quad \& \quad \pi_k \geq 0 \tag{3.59}$$

Figure 3.14 shows a typical PNN structure, where the input neurons are used to accept input data, the kernel functions transfer the multi-dimensional inputs to likelihood values using Eq. (3.55), the empirical model neurons output class likelihood using Eq. (3.53), and the output neuron outputs the decision made based on Eq. (3.58).

The performance evaluation of the PRW model depends on the smoothing parameter or the window size w (or σ^2). Figure 3.15

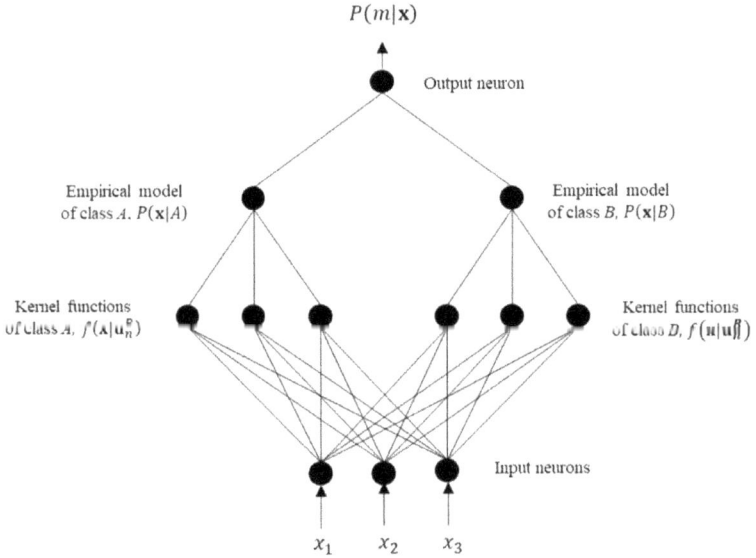

Figure 3.14. The probabilistic neural network structure. It is assumed that there are three inputs (three-dimensional), i.e., $\mathbf{x} = (x_1, x_2, x_3)$. There are two empirical models if there are two classes in data, i.e., $P(\mathbf{x}|A)$ and $P(\mathbf{x}|B)$ as well as two sets of kernel density functions $f(\mathbf{x}|\boldsymbol{u}_n^k)$.

shows a demonstration of the use of PRW for univariate normal distribution. The raw data were composed of 500 data points drawn from a normal distribution. Among these 500 data points, 50 data points were randomly drawn and were used as vectors. The smoothing parameter was set at 0.1, 0.2, 0.3, 0.4, and 0.5. It can be seen that the performance evaluation of the estimated density highly depends on the use of the smoothing parameter.

We now review how PRW can map a nonlinear space to a linear space or a closer to linearly separable space. Figure 3.16(a) shows the XOR data, which are obviously nonlinearly separable. The data are composed of two classes denoted by the triangles (α) and the open dots (β). Four clusters are labeled by four letters, i.e., A, B, C, and D. Figure 3.16(b) shows a PRW model for modeling the XOR data shown in Figure 3.16(a), where the kernel densities are denoted by $f(\mathbf{x}|A)$, $f(\mathbf{x}|B)$, $f(\mathbf{x}|C)$, and $f(\mathbf{x}|D)$. $P(\mathbf{x}|\alpha)$ and $P(\mathbf{x}|\beta)$ are the empirical models (densities). Finally, $P(m|\mathbf{x})$ stands for posterior

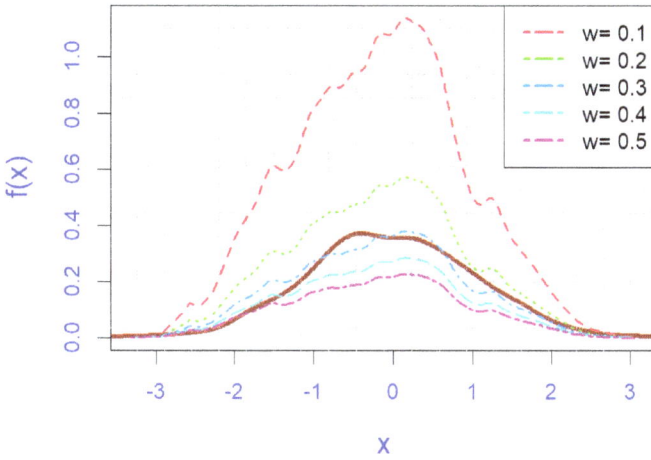

Figure 3.15. An illustration of how PRW works. The solid line represents the raw distribution. The dotted lines represent the estimated densities based on different window sizes.

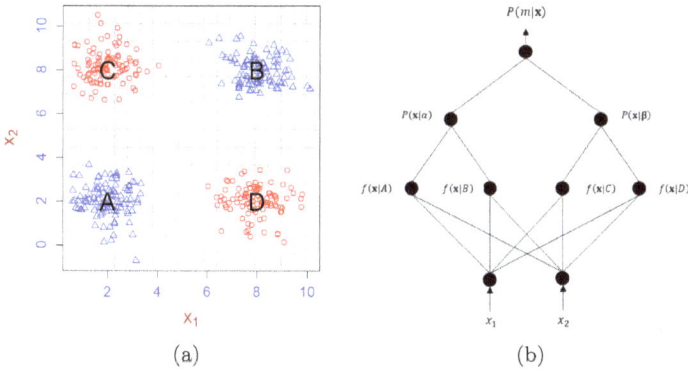

Figure 3.16. The XOR data for analyzing how PRW and the radial basis function algorithm work for the nonlinear data. The triangles and the open dots represent two classes of data. (a) The data. (b) The PRW model constructed for the data.

probability for decision-making, where m is either α or β. Figure 3.17 shows the distance maps for all data points shown in Figure 3.16(a) with regard to the four kernels. It must be noted that these distance maps show that the data are still nonlinearly separable.

Figure 3.18(a) shows the kernel density curves, $f(\mathbf{x}|A)$, $f(\mathbf{x}|B)$, $f(\mathbf{x}|C)$, and $f(\mathbf{x}|D)$. It can be seen that the four kernel density

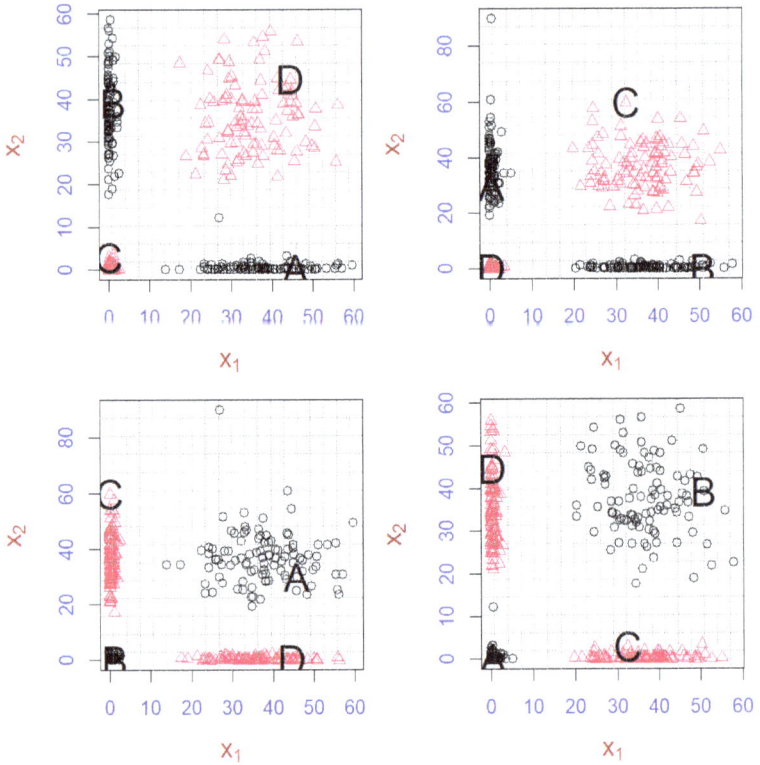

Figure 3.17. The distance maps of the data points (the data shown in Figure 3.16) with regard to four kernels.

curves display very distinct patterns. Finally, Figure 3.18(b) shows the empirical density curves $P(\mathbf{x}|\alpha)$ and $P(\mathbf{x}|\beta)$, where it is obvious that data have been linearly separable. By applying the Bayes rule, posterior probability can be used for decision-making. In this model, it can be seen that the original data are mapped to a higher-dimensional space using the kernels, i.e., a four-dimensional space.

Rather than using a uniform variance, the heteroscedastic PNN was trained using different variances for different kernels so that it can be more robust to model different data (Yang and Chen, 1998; Yang *et al.*, 2000; Venkatesh and Gopal, 2011; Nivethitha *et al.*, 2018).

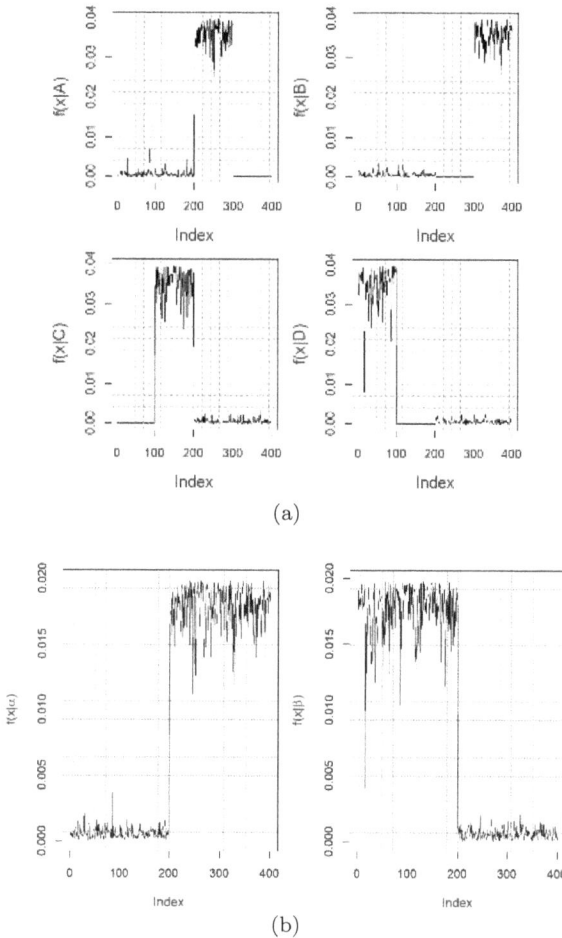

Figure 3.18. Model outputs for the XOR data shown in Figure 3.16(a) using PRW. (a) The kernel densities. (b) The empirical densities.

3.6 Radial basis functions

The previous section discussed the use of more than one kernel vector in PNN (Specht, 1990), where the transfer function is the Gaussian distribution and the outputs are made following probability theory, especially the Bayesian rule. To generalize this kind of kernel machine, radial basis function neural network (RBFNN) is the next important development in this area (Broomhead and Lowe, 1988; Hartman *et al.*, 1990; Bishop, 1991; Chen *et al.*, 1991; Park and

Sandberg, 1991; Musavi *et al.*, 1992; Friedhelm *et al.*, 2001). The radial basis function (RBF) has its foundation in the Euclidean space; that is, the mutual relationship between two data points is quantified using the Euclidean distance shown as follows:

$$\mathcal{K}(\mathbf{x}, \boldsymbol{u}) = \exp(-\varpi \|\mathbf{x} - \boldsymbol{u}\|^2) \tag{3.60}$$

where $\mathbf{x} \in \mathcal{R}^d$ is a data point, $\boldsymbol{u} \in \mathcal{R}^d$ is the Gaussian center of a radial basis, and $\varpi \in \mathcal{R}$ is the smoothing parameter. The RBF matrix for a problem is symmetrical and positively definite if $\mathbf{x}_n = \boldsymbol{u}_n$ and $N = H$:

$$\begin{pmatrix} \mathcal{K}(\mathbf{x}_1, \boldsymbol{u}_1) & \mathcal{K}(\mathbf{x}_1, \boldsymbol{u}_2) & \cdots & \mathcal{K}(\mathbf{x}_1, \boldsymbol{u}_H) \\ \mathcal{K}(\mathbf{x}_2, \boldsymbol{u}_1) & \mathcal{K}(\mathbf{x}_2, \boldsymbol{u}_2) & \cdots & \mathcal{K}(\mathbf{x}_2, \boldsymbol{u}_H) \\ \vdots & \vdots & \ddots & \vdots \\ \mathcal{K}(\mathbf{x}_N, \boldsymbol{u}_1) & \mathcal{K}(\mathbf{x}_N, \boldsymbol{u}_2) & \cdots & \mathcal{K}(\mathbf{x}_N, \boldsymbol{u}_H) \end{pmatrix} \tag{3.61}$$

The RBFNN was proposed using RBF, treating each data point as a center:

$$\hat{y}(\mathbf{x}) = \sum_{h=1}^{H} w_h \mathcal{K}(\mathbf{x}, \boldsymbol{u}_h) \tag{3.62}$$

where $w_h \in \mathcal{R}$ is the coefficient associated with the hth RBF and $\boldsymbol{u}_h \in \mathcal{R}^d$ is the center of the hth radial basis (hypothetical kernel). There are many algorithms for training the RBFNN model; a simple, classical one is the gradient descent approach. We first show how the regression objective function can be used and also assume that we have N data points (training vectors) and H bases (kernels) for constructing the RBFNN model:

$$\varepsilon = \frac{1}{N} \sum_{n=1}^{N} (y_n - \hat{y}_n)^2 + \frac{1}{2} \left(\sum_{h=1}^{H} \lambda_w w_h^2 + \lambda_v v_h^2 \right) \tag{3.63}$$

The estimation of the model parameters (w_h, v_h and $\boldsymbol{u}_h$) uses the following iterative process:

$$\Delta \boldsymbol{u}_h = -\eta \frac{\partial \varepsilon}{\partial \boldsymbol{u}_h}$$

$$\Delta w_h = -\eta \frac{\partial \varepsilon}{\partial w_h}$$

$$\Delta v_h = -\eta \frac{\partial \varepsilon}{\partial v_h} \tag{3.64}$$

where $0 < \eta < 1$. Suppose Eq. (3.62) is replaced by the sigmoid function shown in the following:

$$\hat{y}(\mathbf{x}) = \frac{1}{1 + \exp\left(-\sum_{h=1}^{H} w_h \phi(\mathbf{x}, \boldsymbol{u}_h)\right)} \tag{3.65}$$

In this situation, the maximum likelihood function can also be used for estimating model parameters. The Bernoulli function is first employed:

$$p_n = \hat{y}_n^{y_n} (1 - \hat{y}_n)^{1-y_n} \tag{3.66}$$

where $y_n \in \{0, 1\}$ and $\hat{y}_n \in (0, 1)$. The maximum likelihood function employing the Bernoulli function is thus shown as follows:

$$L = \log \prod_{n=1}^{N} p_n - \frac{1}{2} \left(\sum_{h=1}^{H} \lambda_w w_h^2 + \lambda_v v_h^2 \right) \tag{3.67}$$

Model parameters can also be estimated by maximizing the likelihood function defined in the following:

$$\Delta \boldsymbol{u}_h = -\eta \frac{\partial L}{\partial \boldsymbol{u}_h}$$

$$\Delta w_h = -\eta \frac{\partial L}{\partial w_h} \tag{3.68}$$

$$\Delta v_h = -\eta \frac{\partial L}{\partial v_h}$$

Figure 3.19 show a typical structure of the RBFNN. Figure 3.20 shows the radial basis function neural network model designed for the XOR data shown in Figure 3.16(a), where the output weights have been well designed. It can be seen that the original two-dimensional space is mapped to a four-dimensional space. Figure 3.21 shows the outputs of the RBFNN model constructed for the XOR data shown in Figure 3.16(a). Figure 3.21(a) shows the kernel output which still displays nonlinearity, while the output map shown in Figure 3.21(b) shows that two classes of the data points can be easily separated.

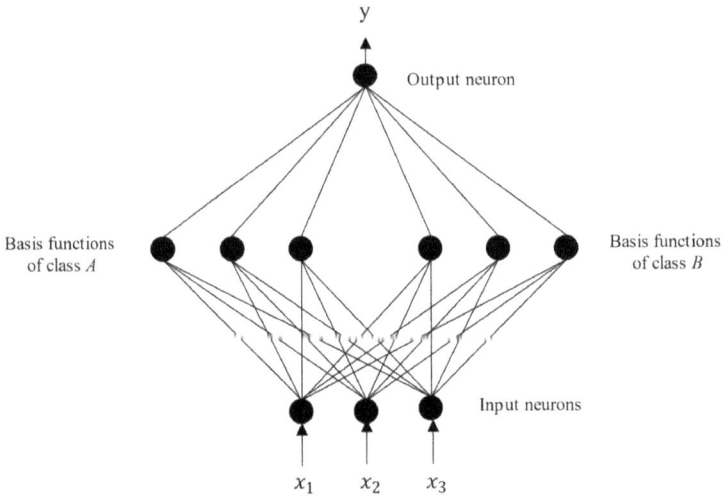

Figure 3.19. The radial basis function neural network structure. It is assumed there are three inputs, i.e., $\mathbf{x} = (x_1, x_2, x_3)$. There are two sets of basis functions for the two classes in data.

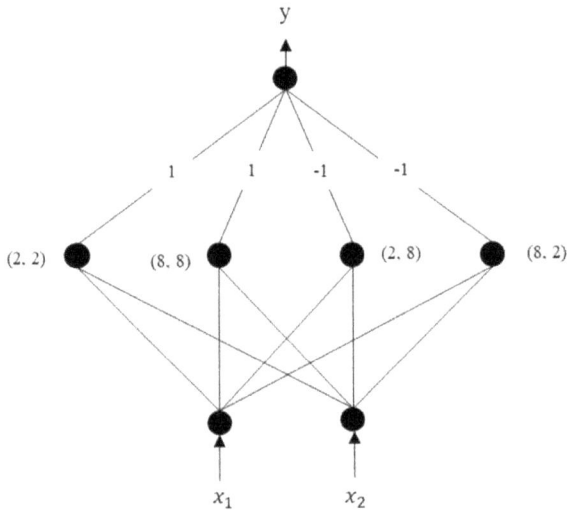

Figure 3.20. The RBFNN model for the XOR data shown in Figure 3.16(a). The centers of four kernels are $(2, 2)$, $(8, 8)$, $(2, 8)$, and $(8, 2)$. The weighs between the hidden neurons (kernels) and output neuron are 1, 1, -1, and -1, respectively.

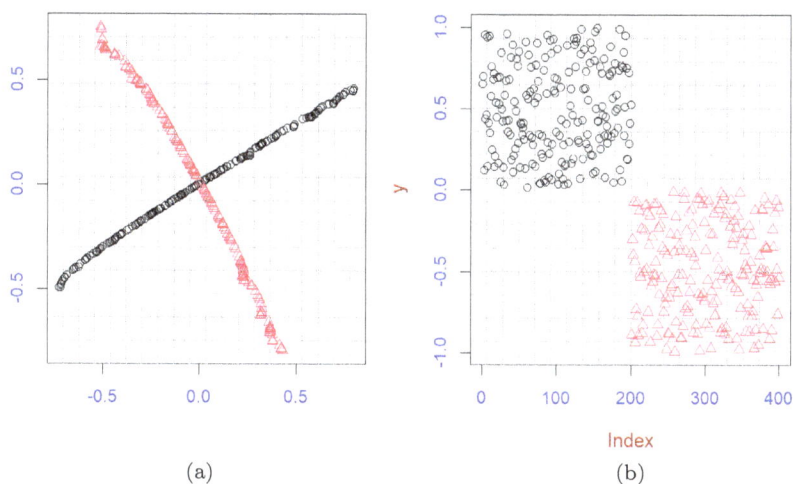

Figure 3.21. The RBFNN model constructed for the XOR data shown in Figure 3.16(a). (a) The Sammon mapping of the kernel outputs of the model. (b) The output map of the model.

There are different algorithms for improving the performance of the RBFNN, such as the orthogonal least-squares algorithm (Chen *et al.*, 1991), the median RBFNN (Bors and Pitas, 1996), the sequential learning model (Lu *et al.*, 1997), the Bayesian learning method (Yang, 2006), the feature selection enhanced algorithm (Chiang and Ho, 2008), and the principal component analysis enhanced algorithm (Ghosh-Dastidar *et al.*, 2008).

3.7 Support vector machine

The basic principle of the support vector machine (SVM) algorithm (Cortes and Vapnik, 1995) is to use a kernel function to map a raw data space to a kernel space. In most situations, the raw data space is nonlinearly separable, but the mapping space is linearly separable or is closer to linearly separable. This is similar to what has been reviewed in the previous section, i.e., RBFNN (Aizerman *et al.*, 1964; Bishop, 2006).

SVM has the advantage of being a simple model that handles complicated data (Hearst *et al.*, 1998). Apart from the rigorous

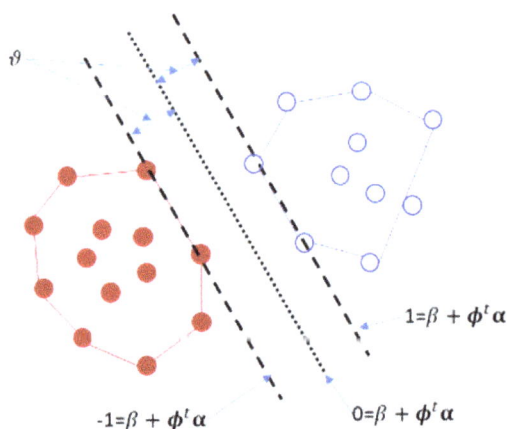

Figure 3.22.   An illustration of the hyperplane used in SVM. The filled dots and the open dots represent two classes of data. The dotted lines represent the optimal hyperplane. The dashed lines represent the margin. The solid lines represent the convex of two classes of data points, while the support vectors on the border of the convex have the shortest distance to the hyperplane.

learning theory, SVM has two major practical issues (Vapnik, 1995; Hearst *et al.*, 1998). The first is that the hyperplane in a SVM model, which separates two classes of data, is always optimal. The second is the so-called kernel trick as discussed earlier. Figure 3.22 shows the working principle of SVM. The optimal hyperplane is satisfied by the following condition:

$$0 = \beta + \phi^t \alpha \qquad (3.69)$$

It maps the data points to a value zero on this optimal hyperplane. Based on this hyperplane, two classes of data points are labeled by either negative or positive values. The support vectors are those data points which have the shortest distance to the optimal hyperplane. The margin which measures the distance between the optimal hyperplane and the support vectors on either side is ϑ. The total margin between two types of support vectors is thus 2ϑ. The relationship between the variables ϕ and a hyperplane $\mathbf{w}$ employed in the SVM model is defined as follows:

$$\hat{y} = \beta + \phi^t \mathbf{w} \qquad (3.70)$$

In SVM, the hyperplane is optimized through quadratic programming so that it can have the maximum margin between two classes of data points. The prime mode of SVM (Cortes and Vapnik, 1995) optimizes the following objective function:

$$\min_{\mathbf{w}} \left\{ \frac{1}{N} \sum_{n=1}^{N} \xi_n + \lambda \|\mathbf{w}\|^2 \right\} \tag{3.71}$$

In the previous notation,

$$\xi_n = \max[0, 1 - y_n(\boldsymbol{\phi}^t \mathbf{w} - \alpha)] \tag{3.72}$$

In order to optimizing the previous equation, it is subjected to the following:

$$y_n(\boldsymbol{\phi}^t \mathbf{w} - \alpha) \geq 1 - \xi_n \tag{3.73}$$

as well as

$$\xi_n \geq 0 \tag{3.74}$$

The dual mode of the SVM model (Cortes and Vapnik, 1995) optimizes the following objective function:

$$\max \left\{ \frac{1}{N} \sum_{n=1}^{N} \alpha_n - \frac{1}{2} \sum_{i=1}^{N} \sum_{j=1}^{N} y_i y_j \alpha_i \alpha_j \boldsymbol{\phi}_i^t \boldsymbol{\phi}_j \right\} \tag{3.75}$$

The previous optimization is subjected to

$$\sum_{n=1}^{N} \alpha_n = 0 \quad \text{and} \quad 0 \leq \alpha_n \leq \frac{1}{2N\lambda} \tag{3.76}$$

Finally, the model parameters used for prediction $\mathbf{w}$ are defined as:

$$\mathbf{w} = \sum_{n=1}^{N} \alpha_n y_n \boldsymbol{\phi}_n \tag{3.77}$$

Note that when $\alpha_n > 0$, it means the corresponding vector will contribute to the prediction using a constructed SVM model. These vectors are referred to as the support vectors. A decision is made using the following formula, where y_n is the class label of the nth

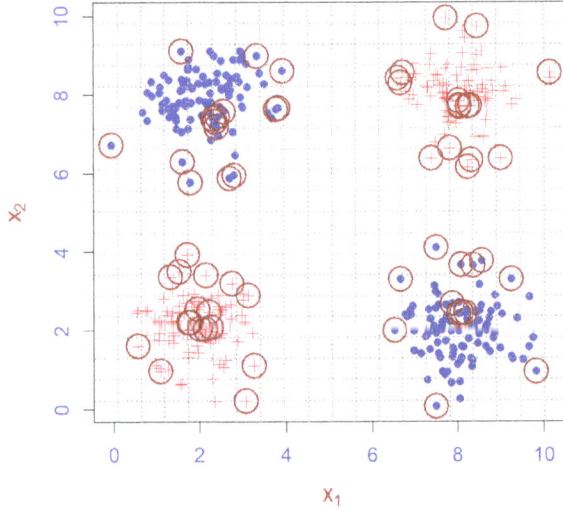

Figure 3.23. An illustration of the support vectors for the data shown in Figure 3.16(a). The filled dots and pluses stand for the two classes of data. The large circles surrounding the small dots and pluses stand for the support vectors.

support vector, $\hat{y}$ is the prediction, and α_n is the positive parameter of the nth support vector determined by SVM:

$$\hat{y} = \text{sign}\left(\sum_{n=1}^{N} \alpha_n y_n \mathcal{K}(\phi, \phi_n)\right) \qquad (3.78)$$

Figure 3.23 shows the output of the SVM model constructed for the data shown in Figure 3.16(a), where it can be seen that support vectors are the data points which are mainly located on the classification boundaries of two classes.

3.8   Relevance vector machine

The relevance vector machine (RVM) algorithm is another kernel machine which employs the Gaussian kernel function (Tipping, 2001). The Gaussian kernel of RVM is shown as follows:

$$p(\mathbf{y}|\mathbf{w}, \beta^{-1}) = \left(\frac{\beta}{2\pi}\right)^{N/2} \exp\left\{-\frac{\beta}{2}(\mathbf{y} - \mathbf{Kw})^t(\mathbf{y} - \mathbf{Kw})\right\} \qquad (3.79)$$

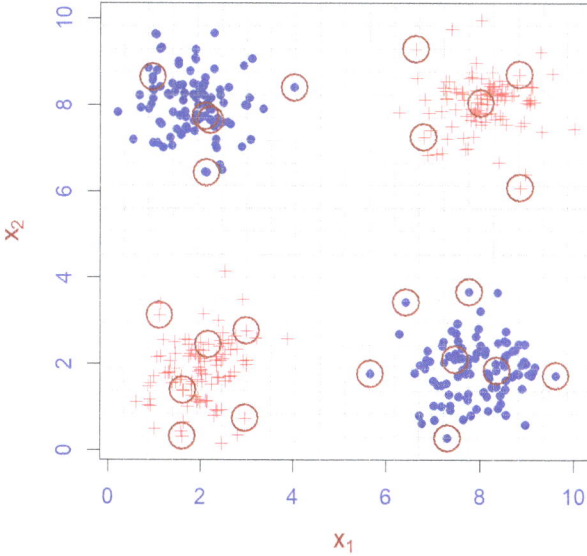

Figure 3.24. An illustration of the relevance vectors for the data shown in Figure 3.16(a). The filled dots and pluses stand for the two classes of data. The large circles surrounding the small dots and pluses stand for the relevance vectors.

In this machine, β is a smoothing parameter, $\mathbf{y}$ is a vector of data labels, $\mathbf{w}$ is a vector of the linear coefficients, and $\mathbf{K}$ is a design matrix, i.e., a kernel space. RVM is trained using the Bayesian learning procedure. During the training process of the RVM model, the *a priori* structure is designed as the linear coefficients given a vector of the hyper-parameters for $\mathbf{w}$:

$$p(\mathbf{w}|\alpha) = \prod_{n=1}^{N} G(w_n|0, \alpha_n^{-1}) \qquad (3.80)$$

The posterior over $\mathbf{w}$ is derived using the Bayes rule, and marginal likelihood is established for the model (Tipping, 2001). Figure 3.24 shows the outcome of the RVM model constructed for the data shown in Figure 3.16(a), where it can be seen that the RVM model employed fewer numbers of relevance vectors than the SVM model constructed for the same dataset.

3.9   Summary

This chapter has discussed how several kernel machines have been developed after introducing several machine learning algorithms based on vector operations. For instance, the Fisher discriminant analysis and the K-means algorithm are classical machine learning algorithms which are based on vector operations, i.e., vector projection and the Euclidean distance between vectors. By kernelizing these classical machine learning algorithms using the kernel trick, their kernelized versions can help analyze complicated data. Through the introduction of different kernel machines, we can understand how the kernel trick plays a key role in kernel machine design and development. This chapter has introduced the earliest kernel machines, such as the Parzen–Rosenblatt windows and the radial basis function neural network. Both show how a feature space is mapped to a kernel space, in which a complicated problem becomes an easy-to-handle problem. Following the introduction of these two earliest kernel machines, this chapter has also introduced cutting-edge kernel machines, including the support vector machine and the relevance vector machine. These two kernel machines have attracted much research interest and many new algorithms based on these cutting-edge kernel machines have been developed and tested in real-world applications with success. Through many real-world applications, these kernel machines have proven to outperform most other machine learning algorithms in both generalization capability and model explanation power. Therefore, the rest of the book will focus on the introduction of bio-kernel machines and the discussion of how they are developed based on these cutting-edge kernel machines to model and analyze protein peptide data.

Chapter 4

Whole Sequence Kernels

In the field of biological pattern discovery, we encounter an issue of how to compare whole DNA or protein sequences for annotating and inferring molecular structure and function for a novel sequence. The comparison between these sequences is conducted through an alignment process to quantify the similarity (or distance) between sequences. Two issues are associated with an alignment process. The first is the measurement metric by which the similarity (or distance) between the residues of sequences under a comparison can be quantitatively and accurately determined. The second is the algorithm by which an alignment can be carried out with satisfactory accuracy and efficiency. Therefore, researchers have made huge efforts in these two areas. This chapter mainly focuses on the algorithmic issue and briefly reviews the sequence homology alignment algorithms followed by the introduction of whole sequence kernel machines.

4.1 Sequence alignment algorithms

Molecular sequences such as DNA sequences or protein sequences are important resources in biology research. For instance, if two sequences share a large degree of similarity, two proteins are very likely to have similar structures as well as similar biological functions (Lipman *et al.*, 1989; Thompson and Plewniak, 1999). Due to this, various algorithms for measuring how two sequences are similar to each other have been developed, such as the Seller algorithm (Sellers, 1974), the Needleman–Wunsch algorithm (Needleman and Wunsch, 1970) and the Smith–Waterman algorithm (Smith and

Waterman, 1981). The first two are global alignment algorithms, and the last one is a local alignment algorithm. But all of them employ the dynamic programming approach in an alignment process. Such an alignment process includes two major stages, i.e., a forward propagation stage followed by a backward propagation stage. During the forward propagation stage, the abovementioned three alignment algorithms employ different metrics for comparison. The Seller algorithm and the Needleman–Wunsch algorithm employ a binary metric when comparing two residues from two sequences while the Smith–Waterman algorithm introduces a gap penalty, hence a non-binary metric. In the forward stage, these alignment algorithms calculate the pairwise edit distance based on a pre-defined distance or similarity metric. Suppose we have two sequences for an alignment, $\mathbf{x} = x_1 x_2 \cdots x_n \in \mathcal{A}^n$ and $\mathbf{y} = y_1 y_2 \cdots y_m \in \mathcal{A}^m$. Two sequences do not necessarily have the same length ($m \neq n$). $\mathcal{A}$ is a set of 20 amino acids for protein sequences or four nucleic acids for DNA sequences. As a typical example, the edit distance used by the Seller algorithm to align these two sequences is shown as follows:

$$d(\mathbf{x}, \mathbf{y}) = d(\tilde{\mathbf{x}}, \tilde{\mathbf{y}})$$

$$= \sum_{i=1}^{N} \delta(\tilde{x}_i, \tilde{y}_i) \tag{4.1}$$

$$= \delta(\tilde{x}_1, \tilde{y}_1) + \delta(\tilde{x}_2, \tilde{y}_2) + \cdots \delta(\tilde{x}_N, \tilde{y}_N)$$

where $N \geq \max(n, m)$ is the length of two aligned sequences $\tilde{\mathbf{x}} = \tilde{x}_1 \tilde{x}_2 \cdots \tilde{x}_N \in \tilde{\mathcal{A}}^N$ and $\tilde{\mathbf{y}} = \tilde{y}_1 \tilde{y}_2 \cdots \tilde{y}_N \in \tilde{\mathcal{A}}^N$. Note that $\tilde{\mathcal{A}} \in \{\tilde{\mathcal{A}}, \emptyset\}$ is a set of acids (amino acids or nucleic acids) plus a gap $\emptyset$, which is called a 'white' nucleic acid or amino acid. There is always a one-to-one correspondence between a pair of residues at the same column (residue) within two aligned sequences, such as the pair $\tilde{x}_i \sim \tilde{y}_i$ shown in the above definition. If a gap is introduced in one aligned sequence at one column (one aligned residue position), we will have the residue alignment outcome as $\{x_i, \emptyset\}$ or $\{\emptyset, y_i\}$. The final alignment score is thus a linear sum of all the residuewise similarities (distances) within the aligned sequences, as shown in the above definition. Finally, $\delta(\tilde{x}_i, \tilde{y}_i)$ stands for the edit distance between

Table 4.1. The alignment result for all cells. H stands for a horizontal move. V stands for a vertical move. D stands for a diagonal move. Adapted from Table 7.5 of Yang (2022a).

		0	*1*	*2*	*3*	*4*	*1*	*2*	*3*	*4*
			Edit distance				Moving direction			
		—	T	G	C	T	T	G	C	T
0	—	0	1	2	3	4				
1	T	1	0	1	2	3	D	H	H	H
2	G	2	1	0	1	2	V	D	H	H
3	G	3	2	1	1	2	V	D	D	H
4	C	4	3	2	1	2	V	V	D	H
5	T	5	4	3	2	1	D	V	V	D

two residues $\tilde{x}_i$ and $\tilde{y}_i$. The value of $\delta(\tilde{x}_i, \tilde{y}_i)$ is binary in the Seller algorithm and the Needleman–Wunsch algorithm, i.e., its value is either zero or one.

Suppose we have two sequences, $x = $ CTAG and $y = $ ACTA. Table 4.1 shows the alignment result using the Seller algorithm for these two sequences. The left panel shows the edit distances calculated in the forward propagation using Eq. (4.1). The right panel shows the moving directions that are used in the backward stage for determining the optimal alignment result. Based on the moving direction information provided by the right panel in this table, the optimal alignment using the Seller algorithm for these two sequences is shown as follows:

```
TGGCT
| |||
T-GCT
```

The Needleman–Wunsch algorithm and the Smith–Waterman algorithm work in a similar way for pairwise alignment between two sequences.

4.2   String kernel

No doubt, the basic alignment algorithms reviewed above have the limitations of low computing speed and large computing memory.

The dynamic programming approach is a time-consuming process, and a dynamic programming table requires a huge memory size for aligning two very long sequences. Although they have played an important role when they were developed for aligning small-size and median-size sequences, they cannot efficiently align long and large-scale sequences, especially multiple sequences efficiently even using modern computers. The most difficult problem is their limitation about prediction efficiency. An efficient learning algorithm must be capable of storing and distributing the learned knowledge in the learning system it creates. Therefore, an efficient prediction system does not need to repeat a learning process that has been used to build up a prediction system when new data come. To achieve this goal, there is no doubt that we have to build up a parametric prediction system or model. This is because it is in the model parameters that we store and distribute the learned knowledge. Suppose a prediction system is denoted by $y = f(\mathbf{x}, \mathbf{w})$, where $\mathbf{x}$ denotes an input, $\mathbf{w}$ denotes model parameters, $f(\mathbf{x}, \mathbf{w})$ denotes a prediction function, and finally, $\mathbf{y}$ denotes the prediction made by this prediction function. A learning process normally employs a dataset with a known outcome. For instance, there might be a set of protein sequences. Part of them belong to one family and the rest belong to the other family, i.e., $\Omega^A = \{\mathbf{s}_1^A, \mathbf{s}_2^A, \ldots, \mathbf{s}_n^A\}$ and $\Omega^B = \{\mathbf{s}_1^B, \mathbf{s}_2^B, \ldots, \mathbf{s}_m^B\}$ as well as $\Omega^A \cap \Omega^B = \emptyset$. How a sequence belongs to either family is what is referred to as the knowledge that is hidden in data. The knowledge must be well learned by a prediction system and must also be well distributed to its parameters. In a prediction process, the stored and distributed learned knowledge is utilized to predict a novel sequence in terms of its family belongingness. However, a sequence alignment system with the support of a databank is not a parametric system. Suppose a new protein has been discovered, aligning it with thousands of sequences stored in a databank using the abovementioned alignment algorithms is tedious and inefficient. Therefore, machine learning-based sequence alignment or analysis algorithms have been widely researched and developed. One of the simplest algorithms is the string kernel.

Suppose a set of sequences with N sequences is denoted by $\boldsymbol{S} = (\mathbf{s}_1, \mathbf{s}_2, \ldots, \mathbf{s}_N)$. In this dataset, a sequence is denoted by a chain, i.e., $\mathbf{s}_n = (s_{n1} s_{n2} \ldots s_{n|\mathbf{s}_n|})$, where $|\mathbf{s}_n|$ is the length of the nth sequence in $\boldsymbol{S}$ and $s_{nm} \in \mathcal{A}$, $\forall m \in [1, |\mathbf{s}_n|]$. The string kernel first generates a feature space by counting the frequency of pre-defined sub-sequences which are referred to as the k-mers within sequences (Leslie *et al.*, 2002, 2004; Vishwanathan and Smola, 2002; Leslie and Kuang, 2004; Kuksa *et al.*, 2008). This is why the string kernel is also referred to as the spectrum kernel. Suppose a set of H contiguous sub-sequences of length R is denoted by $\mathcal{H} = (\omega_1, \omega_2, \ldots, \omega_H)$, where $\omega_h \in \mathcal{A}^R$. The sub-sequence (feature) space is thus defined as follows:

$$\boldsymbol{\Phi}_{\mathcal{H}}^{\text{string}}(\mathbf{s}_n) = \{\phi_{nh}\}_{h=1}^{H} = \{\phi_{\omega_h \in \mathcal{H}}(\mathbf{s}_n)\}_{h=1}^{H} \qquad (4.2)$$

The inner product between the k-mers of two sequences $\mathbf{s}_n \in \mathcal{S}$ and $\mathbf{s}_m \in \mathcal{S}$ generates a string kernel $\mathcal{K}_{sring}(\phi_n, \phi_m)$ shown as follows:

$$\mathcal{K}_{\text{sring}}(\mathbf{s}_n, \mathbf{s}_m, \mathcal{H}) = \boldsymbol{\Phi}_{\mathcal{H}}^{\text{string}}(\mathbf{s}_n) \cdot \boldsymbol{\Phi}_{\mathcal{H}}^{\text{string}}(\mathbf{s}_m) \qquad (4.3)$$

Figure 4.1 shows a sequence alignment comparison between the conventional sequence alignment algorithm, i.e., the Needleman–Wunsch algorithm, and the string kernel approach. To test the computing speed, the DNA sequence length is varied between 1,000 and 10,000. The simulation shows that the string kernel approach demonstrates a significantly fast computing speed compared with the Needleman–Wunsch algorithm; see Figure 4.1(a). Moreover, 500 pairs of DNA sequences with random lengths are generated. The alignment score and the similarity calculated using the string kernel are compared. It can be seen from Figure 4.1(b) that the two approaches generate very high correlated similarity measurements. Figure 4.2 shows the heatmaps of the string kernel matrices generated for the SARS-2 DNA sequences collected from two countries, i.e., Germany and Spain. The sequences were downloaded from GISAID (Elbe and Buckland–Merrett, 2017; Shu and McCauley, 2017; Khare *et al.*, 2021). 1908 SARS-2 sequences were from Germany and 5614

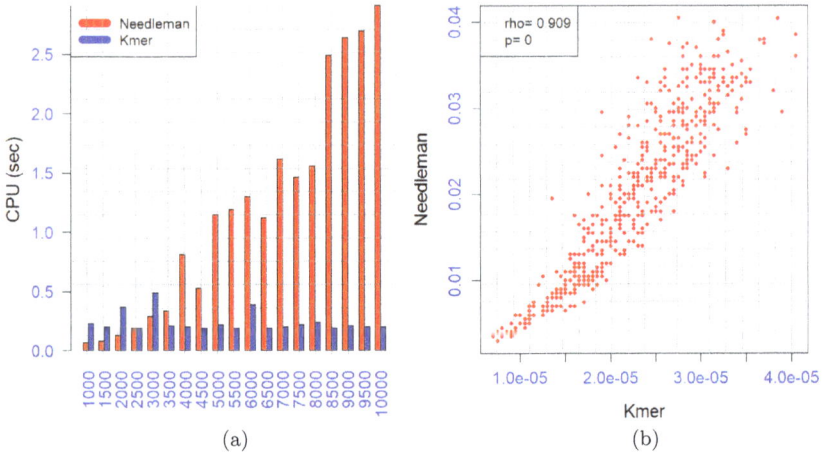

(a)   (b)

Figure 4.1. The comparison between the sequence alignment approach (the Needleman–Wunsch algorithm) and the string kernel approach for comparing sequences in terms of speed and accuracy: (a) *CPU time*: the horizontal axis stands for the sequence length; (b) *accuracy*: 'rho' stands for the correlation coefficient and 'p' is the correlation test p value. The plot is adapted from Figure 7.14 of Yang (2022a).

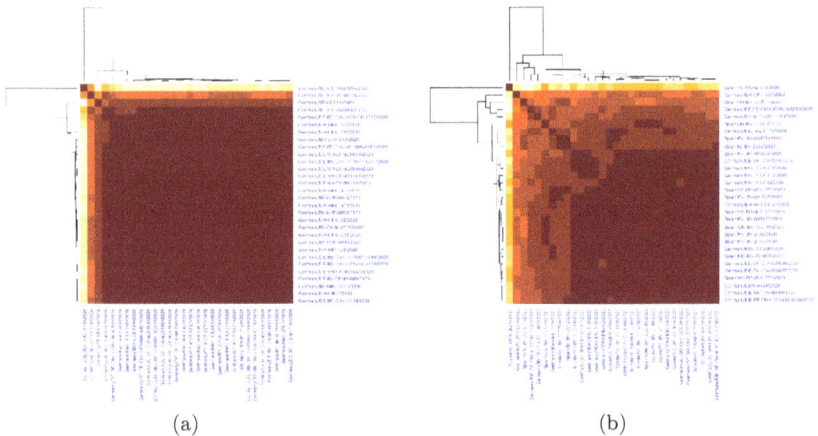

(a)   (b)

Figure 4.2. The heatmaps of the similarity matrices generated by the string kernel for the SARS-2 DNA sequences collected from Germany and Spain: (a) the heatmap of the string kernel matrix generated based on the 30 randomly drawn Germany sequences; (b) the heatmap of the string kernel matrix generated based on the 15 randomly drawn Germany sequences and 15 randomly drawn Spain sequences. The string kernel matrices were generated using the KeBABS package (Palme *et al.*, 2015).

Table 4.2. The confusion matrix of an SVM model constructed on the 7522 SARS-2 sequences using the string kernel.

	Germany	Spain	
Germany	538	34	94.06%
Spain	580	1105	65.58%
	48.12%	97.01%	72.80%

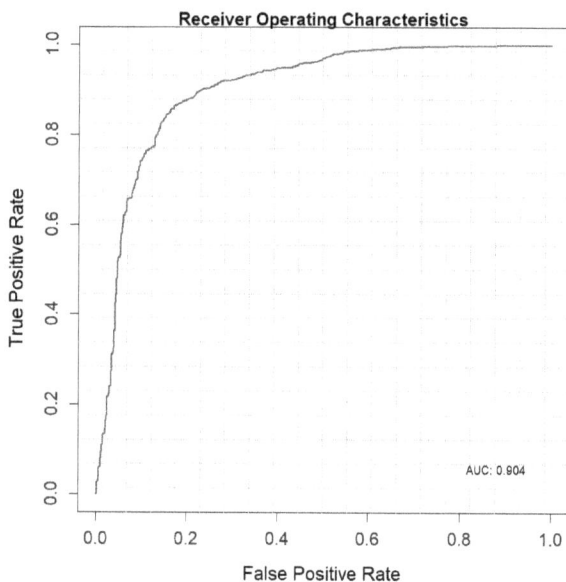

Figure 4.3. The ROC curve for the SVM model constructed on the SARS-2 sequences from Germany and Spain using the string kernel.

SARS-2 sequences were from Spain. The heatmaps show that there is some difference between Germany and Spain.

Based on all these 7522 SARS-2 sequences from two countries, an SVM model was constructed based on the string kernel matrix generated from these sequences. Among 7522 sequences, 70% was used for training the model, and the rest was used for testing. Table 4.2 shows the confusion matrix of such a model. The prediction accuracy for Germany sequences was 94.06% and the prediction accuracy for Spain sequences was 65.58%. The total accuracy was 72.80%. Figure 4.3 shows the ROC curve of the model. The AUC (area under ROC curve) value is 0.904. The ROC curve is normally used for testing the robustness of a classification model, while AUC is used as a quantitative statistic for the robustness measurement of a classification model (Hanley and McNeil, 1982).

4.3 Mismatch kernel

In order to increase the model specificity, the exact string kernel as introduced above has been extended to the mismatch kernel, where a certain number of mismatches between the k-mers in comparison is allowed. The mismatch feature (Leslie *et al.*, 2004) is defined in the following, where $m \ll k$ (k is the length of the k-mers) denotes the maximum allowance of mismatches in a k-mer:

$$\mathbf{\Phi}_{\mathcal{H}}^{\text{mismatch}}(\mathbf{s}_n, m) = \{\phi_{\omega_h \in \mathcal{H}}(\mathbf{s}_n, m)\}_{h=1}^H \tag{4.4}$$

The mismatch kernel is defined as follows:

$$\mathcal{K}_{\text{mismatch}}(\mathbf{s}_n, \mathbf{s}_m, \mathcal{H}) = \mathbf{\Phi}_{\mathcal{H}}^{\text{mismatch}}(\mathbf{s}_n) \cdot \mathbf{\Phi}_{\mathcal{H}}^{\text{mismatch}}(\mathbf{s}_m) \tag{4.5}$$

Figure 4.4 shows the heatmap of the mismatch string kernel matrix generated based on the randomly drawn Germany SARS-2 sequences. This heatmap shows a slight difference from Figure 4.2(a).

4.4 Motif kernel

The motif kernel is another useful kernel function for whole sequence comparison and works based on a motif list defined in advance (Ben-Hur and Brutlag, 2003). A list of discrete sequence motifs is generated using other algorithms such as the eMOTIF, as described in previous research works (Nevill-Manning *et al.*, 1998; Huang and Brutlag, 2001). The use of the motif kernel approach allows more flexibility in a feature extraction process. For an individual sequence $\mathbf{s}_n$, the motif feature can be defined as a vector defined in the following, where $\mathcal{H}$ is a motif set, $|\mathcal{H}|$ is the number of motifs in the motif set $\mathcal{H}$, and ω_h is the hth motif in $\mathcal{H}$:

$$\mathbf{\Phi}_{\mathcal{H}}(\mathbf{s}_n) = \{\phi_{\omega_h \in \mathcal{H}}(\mathbf{s}_n)\}_{h=1}^{|\mathcal{H}|} \tag{4.6}$$

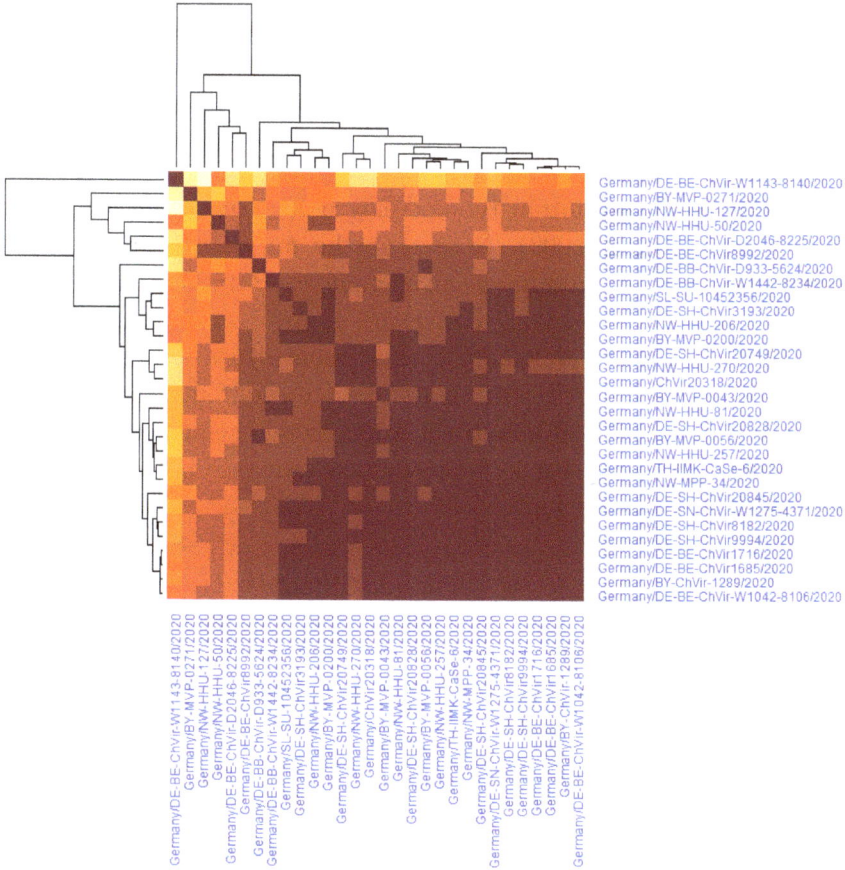

Figure 4.4. The heatmap of the mismatch kernel matrix generated based on the 30 randomly drawn Germany SARS-2 sequences:The mismatch kernel matrix was generated using the KeBABS package (Palme *et al.*, 2015).

The motif kernel is defined as follows:

$$\mathcal{K}_{\text{motif}}(\mathbf{s}_n, \mathbf{s}_m, H) = \boldsymbol{\Phi}_{\mathcal{H}}^{\text{motif}}(\mathbf{s}_n) \cdot \boldsymbol{\Phi}_{\mathcal{H}}^{\text{motif}}(\mathbf{s}_m) \qquad (4.7)$$

4.5 Gappy kernel

The gappy kernel was first proposed for text processing, where the approach was termed as the gappy n-gram kernel (Lodhi *et al.*, 2002) and later developed for analyzing molecular sequences (Leslie and

Kuang, 2003). Mutations, indels, and reading errors in sequencing processes are three commonly encountered problems that cause difficulty in sequence comparison. The gappy kernel was introduced in addition to the mismatch kernel to overcome the sequence comparison difficulty due to residue mutation in sequences caused by the molecular evolution though they belong to the same species (Leslie and Kuang, 2003). Note that the mismatch kernel allows mismatch rather than gaps in an alignment. But the gappy kernel allows gaps in comparison. Using the gappy kernel, a number of gaps (g) are allowed to match a subsequence to a k-mer. We still define $\mathcal{A}$ an alphabet set of characters, i.e., four nucleic acids for DNA sequences and 20 amino acids for protein sequences. Given a subsequence $h \in \mathcal{A}^g$, a (g, k) gappy feature is defined in the following, where $\omega_h \in \mathcal{A}^k$ is one k-mer:

$$\mathbf{\Phi}_{g,k}^{\text{gappy}}(h) = \{\phi_{\omega_h \in \mathcal{A}^k}(h)\}_{h=1}^{|\mathcal{A}^k|} \qquad (4.8)$$

Given a sequence $\mathbf{s}_n$, the gappy feature is thus defined as follows:

$$\mathbf{\Phi}_{g,k}^{\text{gappy}}(\mathbf{s}_n) = \sum_{h \in s_n} \mathbf{\Phi}_{g,k}^{\text{gappy}}(h) \qquad (4.9)$$

The inner product is finally used to generate a gappy kernel:

$$\mathscr{K}_{\text{gappy}}(\mathbf{s}_n, \mathbf{s}_m, \mathscr{H}) = \mathbf{\Phi}_{g,k}^{\text{gappy}}(\mathbf{s}_n) \cdot \mathbf{\Phi}_{g,k}^{\text{gappy}}(\mathbf{s}_m) \qquad (4.10)$$

Figure 4.5 shows the heatmap of the gappy kernel matrix generated based on the 30 randomly drawn Germany SARS-2 sequences.

4.6   Wildcard kernel

In order to allow more flexibility, the wildcard kernel introduces a wildcard character $*$ into the character set $\mathcal{A}$, i.e., $\mathcal{A}^+ = \mathcal{A} \cup \{*\}$.

4.7   Profile string kernel

The profile string kernel introduces probabilities into a feature function (Kuang *et al.*, 2005). Suppose a sequence is denoted by $\mathbf{s}_n = s_{n1}s_{n2}\ldots s_{nm}$, where $m = |\mathbf{s}_n|$, $s_{ni} \in \mathcal{A}$. The probability for

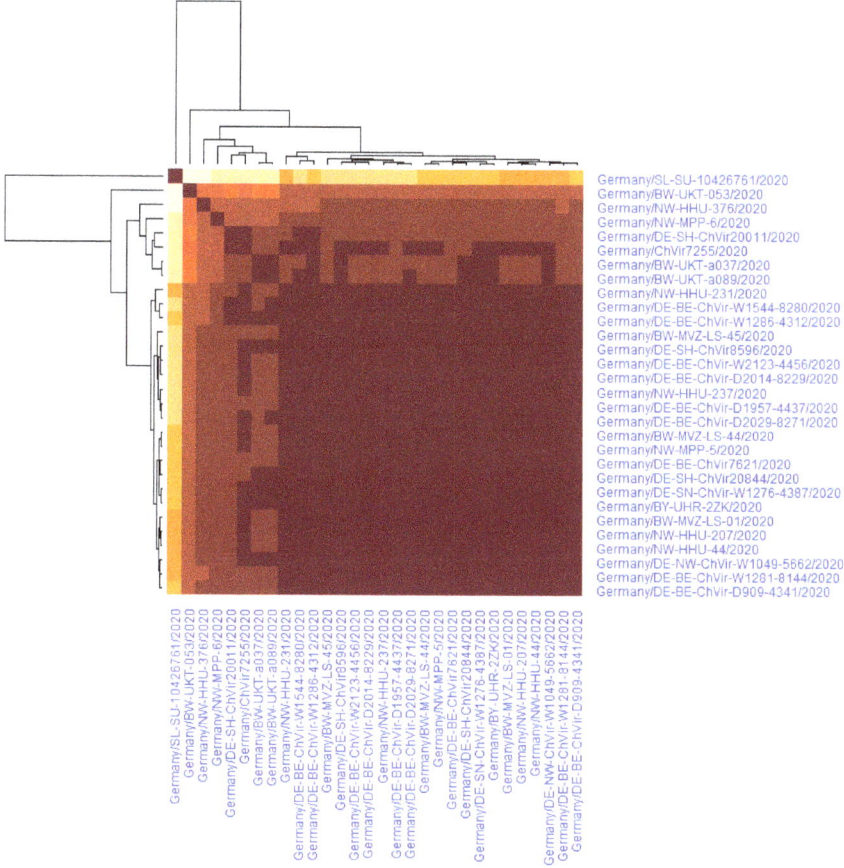

Figure 4.5. The heatmap of the gappy kernel matrix generated based on the 30 randomly drawn Germany SARS-2 DNA sequences: The gappy kernel matrices were generated using the KeBABS package (Palme *et al.*, 2015).

the emission of amino acid α at the ith position of $\mathbf{s}_n$ is defined as $\phi^{\text{profile}}(s_{\text{ni}} = \alpha \in \mathscr{A})$. The profile feature of $\mathbf{s}_n$ is defined as follows:

$$\mathbf{\Phi}^{\text{profile}}(\mathbf{s}_n) = \{\phi^{\text{profile}}(s_{\text{ni}} = \alpha \in \mathscr{A})\}_{i=1}^m \qquad (4.11)$$

The constraint is

$$\sum_{\alpha \in \mathscr{A}} \phi^{\text{profile}}(s_{ni} = \alpha) = 1 \qquad (4.12)$$

The profile string kernel is thus defined as follows:

$$\mathcal{K}_{\text{profile}}(\mathbf{s}_n, \mathbf{s}_m) = \mathbf{\Phi}^{\text{profile}}(\mathbf{s}_n) \cdot \mathbf{\Phi}^{\text{profile}}(\mathbf{s}_m) \qquad (4.13)$$

4.8 Summary

This chapter has introduced several kernel functions (machines) for aligning and comparing whole sequences. Therefore, they are named as whole sequence kernels in this book. These kernel functions, compared with classical sequence alignment algorithms, do improve the efficiency in two ways. First, they are efficient because they don't need to use the dynamic programming approach, which is time-consuming and computing-memory-inefficient. Besides, the second and most important one is that most sequence alignment models generated by sequence alignment algorithms are not parametric prediction systems. However, the kernel functions introduced in this chapter have the capability of generating a parametric prediction model. Such a model can be used for the function prediction of a newly discovered molecule (DNA or protein) based on the pattern stored in its sequence. However, these kernel functions are still frequency-based, i.e., the frequency of k-mers to appear within sequences. Such a spectrum pattern-based approach is in fact an extension to the edit distance, which does not consider the probabilities of the mutual mutations between nucleic acids or amino acids in sequences. Nevertheless, these whole sequence kernel machines have laid a cornerstone for the bio-kernel machines, which will be introduced and discussed in later chapters.

Chapter 5

Mutation Matrix

It is understood that molecules evolve through generations, and it is a constant process in all living organisms and species. If evolution is treated as a phenotypic phenomenon, the genotypic cause of evolution is the mutation between nucleic acids or amino acids. The importance of a mutation matrix in sequence alignment and analysis is that it reflects the true and natural relationship between residues in sequences in terms of molecular evolution. The use of a mutation matrix in sequence alignment can make an alignment result more biologically sound. Based on this knowledge, bio-kernel machines have been developed for modeling and analyzing peptide data, where a peptide, as studied in this book, is a short protein sequence segment in which a functional site (protease cleavage site or post-translational modification site) may be present. This chapter therefore briefly introduces what a mutation matrix is and how a mutation matrix is estimated under a specific mechanism before introducing the bio-kernel machine in following chapters.

5.1 Residue mutation in sequences

The evolution of species or living organismsis is due to sequence mutation, i.e., substitutions between nucleic acids or amino acids. Under evolutionary stress, a nucleic acid or an amino acid at a residue of a sequence may be substituted for another nucleic acid or amino acid from generation to generation. This means that two DNA or protein sequences in two generations will show differences. Several residues of an ancestor sequence will be substituted for new ones in a descendant sequence (Easteal and Collet, 1994; Mirsky *et al.*, 2015;

Yu *et al.*, 2015). The substitution activity also includes insertions and deletions.

It has been shown in research that different nucleic acids or amino acids have different rates or tendencies to substitute under different stresses (Eduardo and Danchin, 2004; Rubinstein *et al.*, 2013; Goldstein and Pollock, 2017; Kende *et al.*, 2023). For instance, an arginine residue tends to mutate to another hydrophilic residue, including glutamine (Taillon-Miller and Shreffler, 1988). The stress also comes from the requirement of molecular structure evolution. The function of a protein is highly dependent on the protein structure, which is in turn highly dependent on the composition pattern of amino acids. The structure of a protein is composed of its local structures. Different local protein structures may require different amino acids because each amino acid is an organic chemical with its own structure and biochemical property (Guzzo, 1965; Atassi, 1966; Schiffer and Edmundson, 1968).

5.2   Substitution/mutation rate

The substitution or mutation rates of nucleic acids or amino acids can be computationally estimated using different approaches (Yang and Nielsen, 2000; Briscoe, 2001; Buzon *et al.*, 2008; Hilton and Bloom, 2018).

The procedure to estimate the substitution rates normally follows three steps in a maximum likelihood process (Felsenstein, 1981; Muse, 1996; Yang and Nielsen, 2000). The first step is to search for the synonymous (silent) and non-synonymous (replacement) sites from two sequences. In the second step, the synonymous and non-synonymous differences between the two sequences are calculated. In this step, the information of the codons which are paired together with the most parsimonious pathways of evolution is employed. In the third step, the substitutions at each site found in the previous steps are adjusted using the standard evolution models (Jukes and Cantor, 1969; Kimura, 1980; Felsenstein, 1981; Nei and Gojobori, 1986; Goldman and Yang, 1994; Yang and Nielsen, 2000). For instance, in Felsenstein's maximum likelihood model (Felsenstein, 1981), two parameters (synonymous α and non-synonymous β) are estimated using a likelihood maximization procedure. The transition/ transversion rate ratio (κ), the synonymous substitution rate d_S, and nonsynonymous substitution rate d_N are finally estimated (Yang and Nielsen, 2000).

5.3 Substitution/mutation matrix

Having understood that nucleic acids or amino acids can mutate through evolution with different rates, encapsulating the substitution rates of nucleic acids or amino acids into a scoring matrix or a mutation matrix can thus be used for further studies. For instance, a mutation matrix can be used to identify the function of a novel protein through sequence alignment (Trivedi and Nagarajaram, 2019).

The Dayhoff mutation matrix is the earliest amino acid mutation matrix, which was developed for sequence comparison through a sequence alignment process (Dayhoff and Schwartz, 1978). It was developed based on 71 families of closely related proteins, in which 1572 mutations were detected. The selection of these sequences was based on two criteria. The first is that they came from the same predecessors and the second is that the alignment similarity between them was greater than 85%. The approach used for generating or estimating the Dayhoff mutation matrix is called accepted point mutation or point accepted mutation (PAM). PAM represents the principle that amino acid substitution in a protein sequence is accepted by the natural selection process. This means that mutations caused by processes other than evolution are excluded in the construction of a mutation matrix in the Dayhoff model.

The Dayhoff mutation matrix or most other similar amino acid mutation matrices usually have 20 rows and 20 columns for 20 amino acids. Each row and each column of such a mutation matrix is associated with an amino acid. In such a matrix, each entry indexed by the nth row and the mth column stands for the substitution likelihood from the nth amino acid to the mth amino acid. During the estimation or construction of the Dayhoff mutation matrix, it is

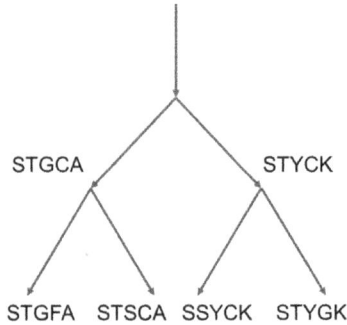

Figure 5.1. An illustration of how accepted point mutation matrix is estimated by the use of a phylogenetic tree.

assumed that the mutation rate from amino acid α to amino acid β is the same as the mutation rate from amino acid β to amino acid α.

Figure 5.1 shows a very simple and naïve example of how Dayhoff and her colleagues generated the mutation matrix. An ancestry sequence was evolved to four offspring through two generations as shown in Figure 5.1. In the first evolution, the ancestor was evolved to two sequences, i.e., STGCA and STYCK, where two mutations are observed. They are between G and Y as well as between A and K. In the next evolution, four mutations were observed: They are the C–F, G–S, T–S, and C–G mutations.

Based on these observations, an accepted point mutation matrix can be formulated as shown in Table 5.1. This table has omitted other amino acids which do not occur in this simplified illustrative example. A real analysis will have many sequences. Therefore, most entries will have numbers which are much greater than one.

Figure 5.2 shows the density of the alignment scores for the alignments between 107 SARS spike glycoprotein sequences downloaded from NCBI. There are $107 \times 106/2 = 5671$ alignments in total. The density shows that the alignment scores are extremely bimodal, meaning that a small subset of sequences have limited mutations that occurred within these sequences, while a majority of sequences have a large degree of dissimilarity.

Table 5.1. An accepted point mutation matrix generated from the observations in the phylogenetic tree shown in Figure 5.1.

	A	C	F	G	K	S	T	Y
A					1			
C			1	1				
F		1						
G		1				1		1
K	1							
S				1			1	
T						1		
Y				1				

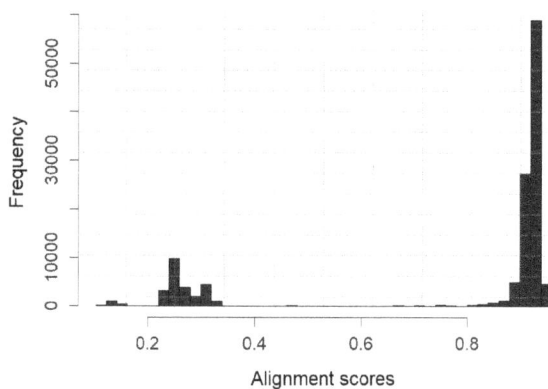

Figure 5.2. The density (histogram) of the alignment scores for the alignment between 107 SARS-2 spike glycoprotein sequences.

Figure 5.3 shows a heatmap of the accepted point mutation matrix constructed using the abovementioned method for these 107 SARS spike glycoprotein sequences. The entries were log-transformed for better visualization. The Smith–Waterman algorithm (Smith and Waterman, 1981) was used to align every pair of two sequences from these 107 sequences. Based on the alignment result, pairwise examination was carried out to detect the mutations across sequences. The total number of heteroscedastic mutations

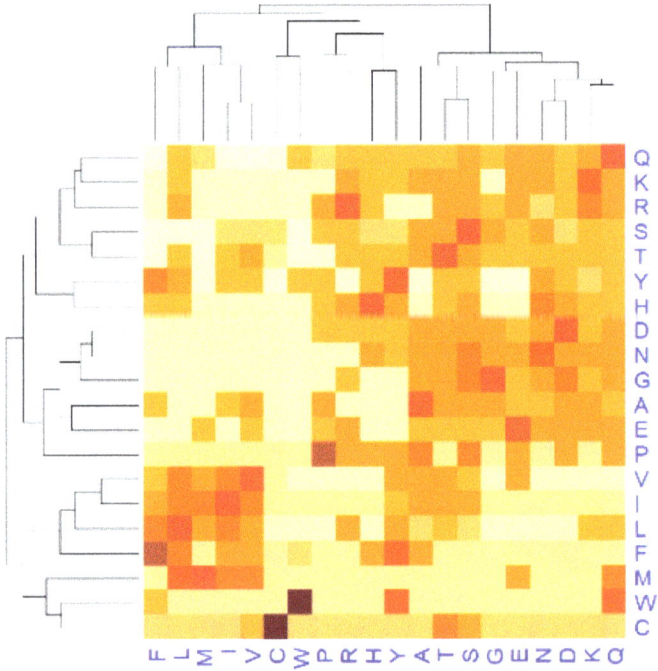

Figure 5.3. A heatmap of the accepted point mutation matrix constructed using the approach described in Dayhoff and Schwartz (1978) for 107 SARS spike glycoprotein sequences. Note that the data were log-transformed for better visualization.

was 391,234. A heteroscedastic mutation means a substitution between two different amino acids. The next issue is to obtain the relative mutability of each of 20 amino acids. Its estimation relies on the calculation of two counts: the frequency of each amino acid (π) and the number by which an amino acid has been substituted (τ). The relative mutability of an amino acid is defined as follows:

$$\sigma_i = \frac{\tau_i}{\pi_i} \quad \forall i \in [1, 20] \tag{5.1}$$

Table 5.2 shows the estimated relative mutability for 20 amino acids in the dataset of 107 SARS spike glycoprotein sequences.

Table 5.2. The estimated relative mutability of 20 amino acids in the dataset of 107 SARS spike glycoprotein sequences. The estimation was done using the Dayhoff approach.

	Observe	Changed	Unchanged	Mutability
A	822355	24119	386393	0.029329
R	433117	13504	202449	0.031179
N	912791	29938	430082	0.032798
D	614527	17240	289750	0.028054
C	328990	125	164122	0.000380
Q	669519	20622	313243	0.030801
E	495507	15085	233529	0.030444
G	906281	13933	438364	0.015374
H	232290	7337	108427	0.031586
I	740159	30769	341490	0.041571
L	1117997	34220	523867	0.030608
K	574691	21504	266624	0.037418
M	130389	6747	56652	0.051745
F	793886	21445	377217	0.027013
P	647131	8466	314266	0.013082
S	1064059	38530	490337	0.036210
T	989013	35126	461003	0.035516
W	100756	435	49550	0.004317
Y	568566	18819	262453	0.033099
V	979812	33270	459866	0.033955

Suppose the accepted point mutation matrix (shown as the heatmap in Figure 5.2) derived from sequences is denoted as $\mathscr{P}$. Each off-diagonal entry of a mutation probability matrix is defined as follows, where λ is a proportionality constant:

$$m_{ij} = \frac{\lambda \sigma_i p_{ij}}{\sum_i p_{ij}} \tag{5.2}$$

Each diagonal entry is defined as follows:

$$m_{ii} = 1 - \lambda \sigma_i \tag{5.3}$$

Figure 5.4 shows the 1 PAM matrix ($\lambda = 1$) for the 107 SARS spike glycoprotein sequences using the Dayhoff approach.

	A	R	N	D	C	Q	E	G	H	I	L	K	M	F	P	S	T	W	Y	V
A	0.971	0	0.001	0.001	0	0.001	0	0.003	0	0.001	0	0.001	0	0	0.007	0.007	0.002	0	0	0.003
R	0	0.969	0	0	0	0.002	0.003	0.001	0.001	0	0.003	0.01	0	0	0.001	0.004	0.003	0	0	0
N	0	0	0.967	0.009	0	0.005	0.002	0.003	0.004	0	0	0.003	0	0	0	0.012	0.002	0	0.001	0
D	0	0	0	0.972	0	0.005	0.003	0.007	0.002	0	0	0.002	0	0	0.002	0.004	0.005	0	0.001	0
C	0	0	0	0	1	0	0	0	0	0	0	0	0	0	0	0	0	0	0	0
Q	0	0	0	0	0	0.969	0.007	0.002	0.001	0	0.001	0.007	0	0	0.006	0.004	0.001	0.001	0.001	0
E	0	0	0	0	0	0	0.97	0.004	0	0	0.012	0.003	0	0	0.003	0.002	0.003	0	0	0.005
G	0	0	0	0	0	0	0	0.985	0	0	0	0	0	0	0	0.011	0.002	0	0	0
H	0	0	0.001	0	0	0	0	0	0.968	0	0.002	0.002	0	0.002	0.001	0.007	0.002	0	0.014	0
I	0	0	0	0	0	0	0	0	0	0.958	0.018	0	0.006	0.002	0	0.001	0.003	0	0	0.018
L	0	0	0	0	0	0	0	0	0	0.001	0.969	0.001	0.007	0.01	0	0	0	0	0.004	0.008
K	0	0.001	0	0	0	0	0	0	0	0	0	0.963	0	0	0.026	0.02	0	0	0.006	0
M	0	0	0	0	0	0.001	0	0	0	0.002	0.002	0	0.948	0	0	0	0	0	0	0.072
F	0	0	0	0	0	0	0	0	0	0	0	0	0	0.973	0	0	0	0	0.037	0.001
P	0	0	0	0	0	0	0	0	0	0	0	0	0	0	0.987	0.008	0	0	0.005	0
S	0	0	0	0	0	0	0	0	0	0	0	0	0	0	0	0.964	0.034	0	0.001	0.001
T	0	0	0	0	0	0	0	0	0	0	0	0	0	0	0	0.001	0.964	0	0.007	0.026
W	0	0	0	0	0	0	0	0	0	0	0	0	0	0	0	0	0	0.996	0.001	0
Y	0	0	0	0	0	0	0	0	0	0	0	0	0	0.001	0	0	0	0	0.967	0.011
V	0	0	0	0	0	0	0	0	0	0.001	0.001	0	0	0	0	0	0	0	0	0.966

Figure 5.4. The 1 PAM generated for the 107 SARS spike glycoprotein sequences using the Dayhoff approach.

5.4 BLOSUM matrix

Unlike PAM, BLOSUM (BLOcks SUbstitution Matrix) clusters the aligned sequences into blocks and derives an amino acid substitution matrix from the blocks (Henikoff and Henikoff, 1992; Eddy, 2004).

Now, ten of the 107 SARS spike glycoprotein sequences are randomly selected to explain how BLOSUM works as shown in Table 5.3. Having aligned them using BLAST (Altschul *et al.*, 1990), Table 5.3 shows one block.

How the substitution score is estimated based on one column of this block is thus illustrated using the Henikoff and Henikoff approach (1992). Suppose the second column is used for the illustration, where nine Ps and one F were observed. Therefore, there are nine PP pairs and one PF pair (a mismatch) for the alignments. A few frequencies or probabilities are required to be derived based on these two figures before deriving a logarithm score ratio as the final substitution score.

The first statistic is the observed occurrence probability of each amino acid pair in a column. This statistic requires the calculation of a few combinations of numbers. Suppose we look at the second

Table 5.3. Ten randomly selected SARS spike glycoprotein sequences for the illustration of BLOSUM matrix construction. The sequence segment shown in this table is one of the blocks.

Accession	Aligned block
8DLL	MPALLSLVSLLS
7R18	MPALLSLVSLLS
7X25	MPALLSLVSLLS
7R15	MPALLSLVSLLS
7R1B	MPALLSLVSLLS
8DLO	MPALLSLVSLLS
6X79	MPALLSLVSLLS
7R19	MPALLSLVSLLS
8DLZ	MPALLSLVSLLS
6VYB	MFVFLVLLPLVS

column of the abovementioned example. The first combination is the number of possible combination pairs, which is 45 ($f_{All} = 10 \times 9/2$). The number of possible matching pairs (PP) is 36 ($f_{PP} = 9 \times 8/2$). The number of possible mismatch pairs (PF) is 9 (f_{PF}). In fact, the relationship between these three observed occurrence probabilities is defined as follows, where Ω is the set of all possible pairs within a column of aligned blocks:

$$f_{All} = \sum_{\omega \in \Omega} f_\omega \tag{5.4}$$

In this example, $f_{All} = f_{PP} + f_{PF}$. More quantities are calculated using the following equation:

$$q_\omega = \frac{f_\omega}{f_{All}} \tag{5.5}$$

Again, the two quantities (only two possible combinations between P and F) are $q_{PP} = 36/45 = 0.8$ and $q_{PF} = 9/45 = 0.2$. Following the abovementioned calculations, we can calculate the occurrence probability for each amino acid within the column. It is defined as follows, where $\tau \in \Omega_1$, Ω_1 is a set of matching pairs, and

Ω_2 is a set of mismatch pairs:

$$p_\alpha = q_{\alpha=\tau} + \frac{1}{2} \sum_{\sigma \in \Omega_2} q_\sigma \tag{5.6}$$

For the example shown earlier, the occurrence probability of P is

$$p_P = q_{PP} + \frac{1}{2} q_{PF} = 0.8 + \frac{1}{2} \times 0.2 = 0.9 \tag{5.7}$$

and the occurrence probability of F is

$$p_F = q_{PF} + \frac{1}{2} q_{PF} = 0 + \frac{1}{2} \times 0.2 = 0.1 \tag{5.8}$$

The expected occurrence probability of a matching pair is defined as follows:

$$\hat{p}_{\alpha\alpha} = q_\alpha^2 \tag{5.9}$$

and

$$\hat{p}_{\alpha\beta} = 2p_\alpha p_\beta \tag{5.10}$$

Thus, we have the following expected occurrence probabilities:

$$\hat{p}_{PP} = q_P^2 = 0.9^2 = 0.81 \tag{5.11}$$

and

$$\hat{p}_{PF} = 2p_P p_F = 2 \times 0.9 \times 0.1 = 0.18 \tag{5.12}$$

Because the sum of all probabilities should be one and the pair FF does not occur in that column, the expected probability of the pair FF is inferred as follows:

$$\hat{p}_{FF} = 1 - \hat{p}_{PP} - \hat{p}_{PF} = 0.01 \tag{5.13}$$

Finally, the base two logarithm applies to ratio of $q_{\alpha\beta}$ over $\hat{p}_{\alpha\beta}$ leading to the substitution ratio as shown in the following, where $[x]$ stands for transferring x to an integer:

$$s_{\alpha\beta} = \left[\log_2 \frac{q_{\alpha\beta}}{\hat{p}_{\alpha\beta}} \right] \tag{5.14}$$

The final substitution (mutation) matrix is completed by repeating the abovementioned procedure across all columns of all blocks (clusters).

5.5 Advanced mutation matrices

After the development of the Dayhoff mutation matrix and the BLOSUM mutation matrix, more advanced mutation matrices were developed. The MDM (mutation data matrix) is one of them (Jones *et al.*, 1992). The estimation of MDM starts from a PAM matrix (Q) and uses a basic matrix, which is a probability matrix of mutations. Each entry of this basic matrix is the probability by which amino acid α mutates to amino acid β if α is located in the ith row and β is located in the jth column of a mutation matrix. It is assumed that the diagonal entries remain unchanged.

Suppose amino acid α is located in the ith row of the matrix. Given the average relative mutability of α (m_i) and a proportionality constant λ, a diagonal entry of the basic matrix is calculated using the following formula:

$$M_{ii} = 1 - \lambda m_i \tag{5.15}$$

An off-diagonal entry of the matrix is calculated as follows:

$$M_{ij} = \frac{\lambda m_i Q_{ij}}{\sum_{j=1}^{20} Q_{ij}} \tag{5.16}$$

In the two abovementioned equations, λ is a key parameter, which is related to the evolutionary distance as seen in the Dayhoff mutation matrix. Similar to PAM, an approximation of the evolutionary distance is denoted by $\mathbb{D}$. The normalized occurrence frequency of an amino acid located in the ith row is denoted by f_i. The following equation is used to define the relationship:

$$\sum_i f_i M_{ii} = 1 - \frac{\mathbb{D}}{100} \tag{5.17}$$

The next stage of MDM construction is to build up a relatedness odds matrix from the basic matrix, which is symmetric. An entry at row i and column j represents the probability of a mutation from amino acid α, which is located in the ith row to amino acid β, which is located in the jth column of the matrix. The entry is defined as follows:

$$P_{ij} = \frac{M_{ij}}{f_i} \tag{5.18}$$

The MDM is finalized using the following equation, where $[x]$ means rounding up x to an integer:

$$S_{ij} - [10\log_{10} F_{ij}] \tag{5.19}$$

The VTML matrix is also an advanced mutation matrix and was developed by evaluating the mutation rates as well as the distances of evolution once a pairwise alignment of sequences has been carried out (Muller and Vingron, 2000). The use of the maximum likelihood as well as the initialization based on the Dayhoff mutation matrix delivers a more confident alignment quality. The algorithm is run in two steps. Based on a set of input parameters, the evolutionary divergence of an alignment is estimated at first. Then, given the divergence time which is estimated, new parameters are estimated.

5.6   Other mutation matrices

In addition to the abovementioned mutation matrices, many other mutation matrices have been developed. Some were estimated based on physio-chemical properties of the amino acids, such as the Grantham matrix (Grantham, 1974), the Miyata matrix (Miyata *et al.*, 1979), and the Rao matrix (MohanaRao, 1987). Some were estimated based on protein structures, such as side-chain accessibility, amino acid type, hydrogen bond formation, and secondary structure. The typical ones include the Risler matrix (Risler *et al.*, 1988), the Naor matrix (Naor *et al.*, 1996), the Gonnet matrix (Gonnet *et al.*, 1992), the Johnson matrix (Johnson *et al.*, 1993), the Russell matrix (Russell *et al.*, 1997), the Prlic matrix (Prlic *et al.*, 2000), and the Keul matrix (Keul *et al.*, 2017). Generally speaking, it is unknown whether these mutation matrices can deliver better performance for the analysis of a wide range of peptide data because they are estimated based on more specific circumstances for specific problems. This issue will be discussed in later chapters.

5.7 Summary

This chapter has briefly introduced several mutation matrices including the earliest one, i.e., the Dayhoff mutation matrix. The development of various mutation matrices has greatly improved the sequence alignment quality. This is because the mutation scores between nucleic acids or amino acids can truly reflect the nature of the evolution of molecules. In later chapters, the biologically sound property of mutation matrices will be explored to improve the quality of short-sequence (peptide) modeling and analysis. Although peptides are short, their evolutionary nature should be the same as those of long or whole sequences. The pattern of amino acid mutation has been found to tolerate randomness in whole sequences (Palzkill and Botstein, 1992; Guo *et al.*, 2004; Wong *et al.*, 2006; Zhao *et al.*, 2014). Therefore, it is unavoidable to observe amino acid mutation within peptides. However, it is understood that functional peptides (including cleaved peptides and post-translational modified peptides) will maintain the pattern of their primary structures or the composition pattern of the amino acids, and hence secondary structures, during evolution. The employment of a mutation matrix to measure the similarity between peptides can explore the biological nature of peptides, especially the functional peptides. In the following chapters, we will explain how these mutation matrices are employed to develop bio-kernel machines for the purpose of peptide pattern discovery and analysis.

Chapter 6

Bio-Kernel Machines

This chapter will introduce the property of protein peptides (short protein sequences). The data for the peptides include protease cleavage peptides and post-translational modification peptides. It must be noted that the purpose of analyzing protein peptides is different from that of analyzing whole protein sequences. This is not only because of the length difference between whole protein sequences and peptides. The main objective of analyzing whole protein sequences is to understand the protein structure and function annotation. However, the objective of analyzing peptides is to detect whether a peptide contains only one functional site or not. The former examines the protein in a global view, while the latter focuses on a protein at its local site. Importantly, a protein may have more than one protease cleavage site or post-translational modification site. But, every peptide only contains one site. The analysis of a set of peptides is conducted to discover the pattern of a type of functional site from the given peptides, for instance, the pattern of HIV protease cleavage or the pattern of Serine post-translational modification. Therefore, aligning a whole protein sequence against a databank of known sequences is the main approach for global protein function or structure annotation. The main contributors to a functional site will not involve remote residues within a protein sequence. Instead, only a few nearby or local residues contribute to the formation of a functional site within a protein. Predicting or discovering a functional site (a protease cleavage site or a post-translational modification site) within a peptide, which only involves one or two residues, does not require the alignment of a whole sequence at all. Therefore, analyzing peptides requires a different strategy. Because of this, different approaches have been developed for peptide pattern discovery and analysis. This chapter starts from the introduction of the earlier approaches developed for analyzing peptide data, including biology-related approaches and computational approaches. The former uses amino acid descriptors (or physio-chemical descriptors) and the latter is based on computational approaches or statistical approaches. This chapter also discusses the property of peptide alignment matrix. Then, this chapter introduces the two earliest versions of the bio-kernel machines, i.e., the least-squares bio-kernel and Fisher discriminant analysis bio-kernel machines. Throughout the introductions, real data examples are used to demonstrate the model generalization performance of these bio-kernel machines.

6.1   Protease

The development of the bio-kernel machines was based on the recognition of the protease structure. The protease structure is therefore briefly introduced. Within the cell, protein is one of the most important components, which is not in a static state and is always in a dynamic state involving many different activities to maintain the life of the cell and the life of the living species (Taylor *et al.*, 1997). It is protein activities that maintain most types of functions of cells and living organisms.

One of the key functions of a protein is the catalyzation carried out by protein enzymes. Sustaining life relies on cellular metabolic processes and cannot proceed without the catalyzation implemented by enzymes (Taylor *et al.*, 1997; Punekar, 2018). Most cellular reactions, such as cellular signaling, are the result of protein enzyme catalyzation. However, the study of protein functions, such as the enzyme activities, is challenged by the huge number of peptides. Given 20 amino acids, a peptide which is a chain of n amino acids can have 20^n possible combinations in nature. Based on the limited data so far, many fundamental studies have been carried out to examine protein functions, especially enzyme functions in relation to the life-sustaining issues of living species. For instance, a study shows that the proteins NF-kappaB and AP-1 have a function in dehydroepiandrosterone metabolite for the enhancement of TNF-alpha activity (Dulos *et al.*, 2005). How a novel protein with life-sustaining functions in living species can be used to revise the gene regulation has been demonstrated (Digianantonio *et al.*, 2017). Due to the importance of the enzymatic activities, measuring the activities can test the gene expression and metabolite change within the cell, especially involving tumor necrosis (Dowd, 2023).

The main force of the enzymatic activities involves the pro- teases, which are also called peptidases, proteinases, and proteolytic enzymes. Their main function is catalyzation, which breaks down proteins (often called poly-proteins) into smaller protein products. Each of these protein products catalyzed from a larger protein or a poly-protein carries out a specific function. Thus, the main function of a protease is cleavage and a protease can be used for digestion, catabolism, and signaling (Hanson and Marzluf, 1975; Taylor *et al.*, 1997; Greenfield *et al.*, 2020; Hasan *et al.*, 2021). Cleavage occurs within the peptide bonds through hydrolysis and bonds are broken using water. A protease is a protein with a short chain of amino acids. For instance, a Staphylococcal protease is composed of about 250 amino acids (Drapeau, 1978), an HIV protease is composed of about 100 amino acids (Weber *et al.*, 2021), and a SARS-Cov2 main protease is usually composed of about 300 amino acids (Estrada, 2020).

When a protease tries to catalyze or cleave a protein at one site, two proteins will bind together for a chemical reaction. One of them is a protease and the other is a substrate (Griswold *et al.*, 2019; Luo *et al.*, 2019). A protease is normally denoted by

$$\boldsymbol{S} = S_n \ldots S_2 S_1 S_1' \; S_2' \ldots S_m' \tag{6.1}$$

where the cleavage site of the protease $\boldsymbol{S}$ is between S_1 and S_1'. The protease has the length $n + m$, and $S_n \ldots S_2 S_1$ represents the N-terminal part of a protein sequence, while $S_1' S_2' \ldots S_m'$ is the C-terminal part of a protein sequence. Normally, the N-terminal amino acid composition determines the cleavage activity (Luo *et al.*, 2019). The substrate is denoted by

$$\boldsymbol{P} = P_N \ldots P_2 P_1 P_1' \; P_2' \ldots P_C' \tag{6.2}$$

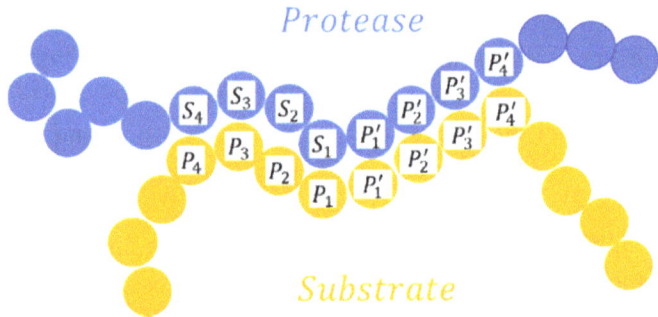

Figure 6.1. An illustration of how a protease and a substrate bind together. The protease sequence is denoted by S and the substrate sequence is denoted by P. The cleavage happens when two proteins bind at the cleavage site, i.e., P_1 and P'_1.

where the cleavage site of the substrate P is between P_1 and P'_1. A cleavage that happens in substrate P leads to two smaller protein products which are labeled $P_N \ldots P_2 P_1$ and $P'_1 P'_2 \ldots P'_C$. In other words, $P_N \ldots P_2 P_1 P'_1 P'_2 \ldots P'_C$ is the original protein sequence (substrate), which is going to be cleaved by the protease S. Note that the whole sequence of the protease can be denoted by $S_\ell \ldots S_2 S_1 S'_1 S'_2 \ldots S'_\imath$, where $\ell + \imath \ll n + m$ and $n + m \neq N + C$. Figure 6.1 illustrates how a protease and a substrate bind together for the cleavage activity. The biochemical reaction of cleavage happens when a protease binds to a substrate. The binding between a protease and a substrate indicates that two proteins have complementary structures at the binding (cleavage) site as well as the neighboring residues. This is an important concept. Suppose one has a concave surface at the binding site. The partner for binding must have a convex surface at the binding site. In addition, the size of the concavity and the size of the convexity of the two proteins must match. This structural requirement means that $S_4 S_3 S_2 S_1 S'_1 S'_2 S'_3 S'_4$ fits $P_4 P_3 P_2 P_1 P'_1 P'_2 P'_3 P'_4$ as shown in the central part in Figure 6.1, while the other parts of the protease and substrate do not need to fit as shown in the two tail parts of Figure 6.1. Note that, in this case, $\ell = \imath = 4$.

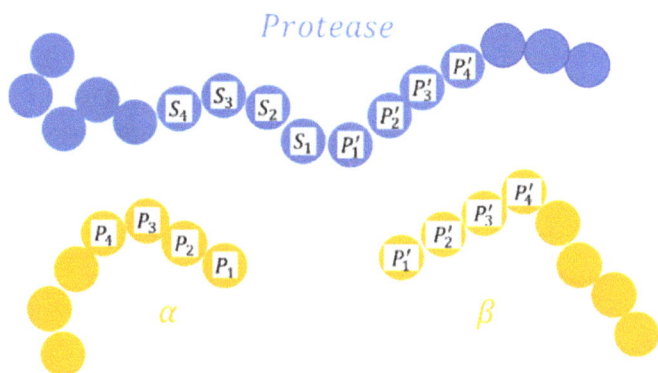

Figure 6.2. The outcome of the protease activity for Figure 6.1. α and β are two protein products which are generated by cleavage.

Figure 6.2 shows the outcome of the protease–substrate binding. The cleavage generates two protein products, i.e., α and β, where $P_4 P_3 P_2 P_1$ is in the product α which is $P_N \ldots P_2 P_1$, while $P'_1 P'_2 P'_3 P'_4$ is in the product β which is $P'_1 P'_2 \ldots P'_C$.

The next important issue is the structural conservation of proteases, especially the structure at the cleavage site of a substrate. It is this property that has led to the development of bio-kernel machines. The property of relatively conserved protein structures has been found to be a fundamental component for protein binding (Branlant *et al.*, 1982; Konc and Janezic, 2007). In the study of protease activities, structural conservation has also been thoroughly verified (Wernerson *et al.*, 2006; Luo *et al.*, 2013). Figure 6.3 shows the protease sequence structure using the sequence logo package (Wagih, 2017) for Factor Xa protease cleavage data (Yang *et al.*, 2006; Yang and Hamer, 2007). The Factor Xa protease sequence is denoted by $P_4 P_3 P_2 P_1 P'_1$, where the cleavage site is between P_1 and P'_1. In this case, $\ell = 4$ and $\imath = 1$. It can be seen that the peptides have a very conserved amino acid pattern. For instance, the amino acid Arginine (single letter notation R) is very conserved at P_1, the amino acid Valine (single letter notation V) is prevailing at P_4, the amino acid Phenylalanine (single letter notation F) and the amino

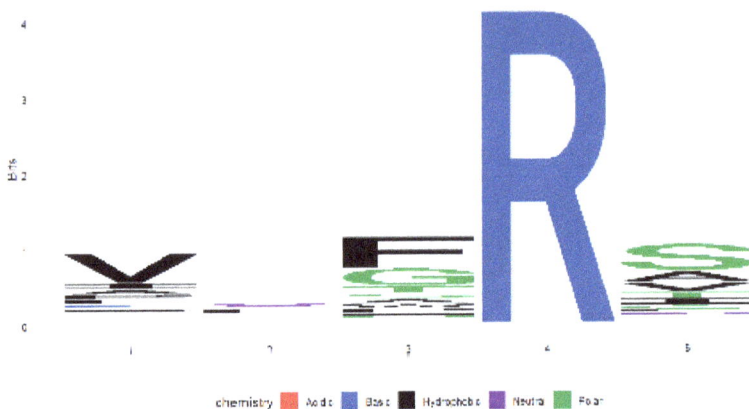

Figure 6.3. The amino acid composition pattern of the cleaved Factor Xa protease substrates. From left to right are the residues P_4, P_3, P_2, P_1, and P'_1.

acid Glycine (single letter notation G) are prevailing at P_2, and the amino acid Serine (single letter notation S) is prevailing at P'_1. This pattern is consistent with the outcome that the N-terminal amino acid composition may dominate the cleavage activity (Luo *et al.*, 2019). It must be noted that the pattern shown in Figure 6.3 does not mean that we can detect a Factor Xa cleaved peptide by looking at the P_1 residue to see whether an Arginine (R) is present because not every Arginine (R) residue is a Factor Xa cleavage site. Only part of the Arginine (R) residues can be cleaved by the Factor Xa protease. The cleavage activity depends not only on the Arginine (R) residue but also on the amino acid composition pattern surrounding the Arginine.

However, the structure beyond the cleavage sites may not be conserved at all. Figure 6.4 shows a sequence logo for non-cleaved peptides for the Factor Xa protease. It can be seen that no site within these peptides shows a significantly conserved pattern of amino acid composition. Importantly, this sequence logo shows that the Arginine (R) is also present in non-cleaved Factor Xa peptides. The key difference between the logo shown in Figure 6.3 and the logo shown in Figure 6.4 is the amino acid composition pattern within two types of the peptides. This is why the prediction of a Factor Xa

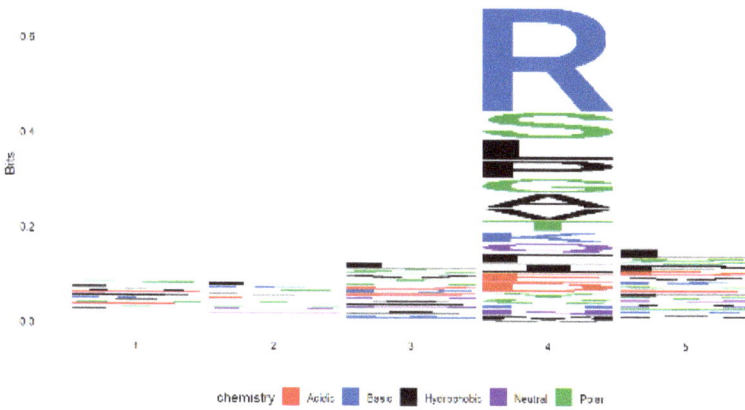

Figure 6.4. The almost random amino acid composition pattern of the non-cleaved Factor Xa protease substrates. From left to right are the residues P_4, P_3, P_2, P_1, and P_1'.

cleaved peptide requires the discovery and analysis of the amino acid composition pattern around an Arginine (R) residue within peptides.

6.2 Post-translational modification

Post-translational modification (PTM) also involves the study of the specificity of the short sequences, and hence the study of peptides. There are many different PTM activities, such as phosphorylation, palmitoylation, acylation, and hydroxylation, to name a few. Each PTM is involved with the change of the properties of a protein through a biosynthesis which adds a chemical component, such as acetyl, phosphoryl, or glycosyl, to the modification site of a protein (Taylor *et al.*, 1997). Unlike a protease cleavage activity which involves the binding of two proteins (protease and substrate), a PTM activity takes place during covalent binding between two atoms when they share electrons (Alam *et al.*, 2008).

PTM plays a key role for signaling within and between the cells. For instance, phosphorylation can be used to control enzyme activity. In a glycosylation, the protein-folding capability can be increased once carbohydrate molecules are attached to a protein. Most PTM sites are functional, meaning that the structure located at PTM

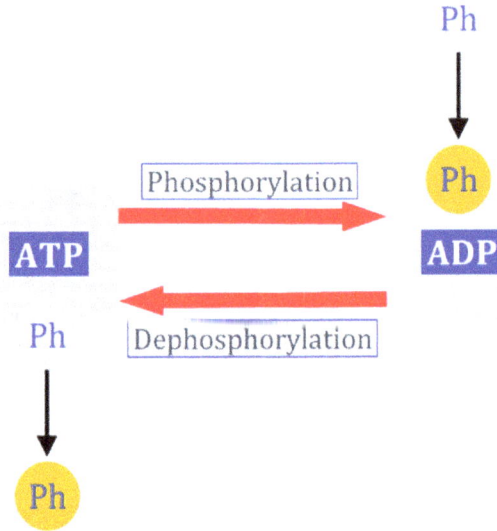

Figure 6.5. An illustration of phosphorylation and dephosphorylation. The phosphorylation process relies on a kinase and will add a chemical phosphor to a protein called adenosine triphosphate (ATP). It results in another protein called adenosine diphosphate (ADP). The dephosphorylation process depends on a phosphatase and will remove the chemical phosphor from a protein called ADP. It leads to ATP.

sites is conserved. Figure 6.5 shows a diagram of phosphorylation and dephosphorylation. In the phosphorylation process, a chemical phosphor is introduced and attached to a protein called adenosine triphosphate (ATP) to transfer it to a new protein celled adenosine diphosphate (ADP) (Rimington, 1927; Macfarlane, 1936; Wiggert and Werkman, 1938). However, a reverse process happens in dephosphorylation, where a phosphor chemical is removed from ADP so that an ADP is transferred back to an ATP. The sequence or a peptide used for a PTM is normally denoted as follows, where P_0 is the modification site:

$$\boldsymbol{P} = P_n \ldots P_2 P_1 P_0 P_1' P_2' \ldots P_n' \tag{6.3}$$

Figure 6.6 shows the sequence logo (Wagih, 2017) for a set of Serine phosphorylated sequences, where P_0 was removed because it is always

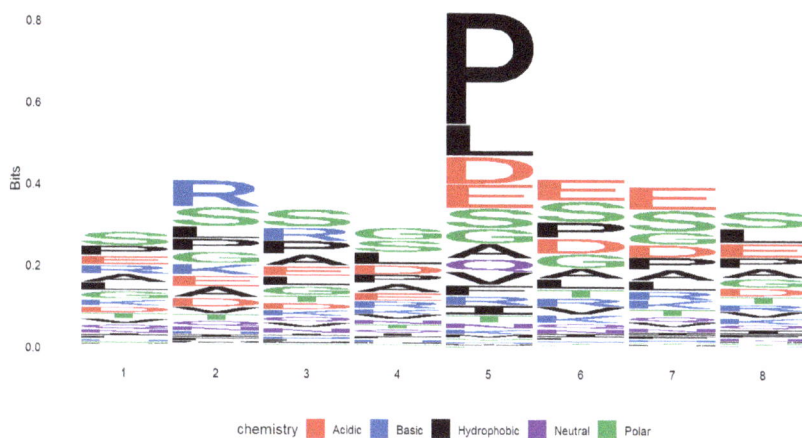

Figure 6.6. The sequence logo of Serine phosphorylated sequences. From left to right are the residues $P_4, P_3, P_2, P_1, P_1', P_2', P_3'$, and P_4'.

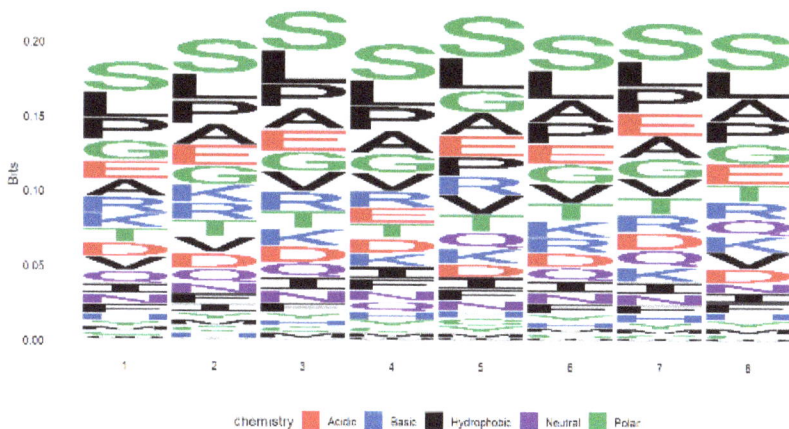

Figure 6.7. The sequence logo of sequences which are not Serine phosphorylated. From left to right are the residues $P_4, P_3, P_2, P_1, P_1', P_2', P_3'$, and P_4'.

occupied by Serine. In the sequence logo, it can be seen that the amino acid Proline (single letter notation P) is prevailing at the site of P_1'. Figure 6.7 shows the sequence logo which is not Serine phosphorylated. It can be seen that there is no conserved pattern at all.

6.3   Early approaches for modeling and analyzing protein peptides

Unlike analyzing whole molecular sequences, where the jobs are done using global or local alignment algorithms, analyzing protein peptides requires a different strategy as discussed earlier. Peptide data have two structural properties, i.e., the length of peptides under study is up to a dozen residues and all peptides within a peptide dataset have the same length. The most important property of peptide pattern discovery is that each peptide may contain the maximum of one functional site (cleavage site of post-translational modification site), and the peptides within a dataset are organized so that the function site under investigation is always located at the fixed position within the peptides, such as $P_1 - P_1'$ or P_0.

However, like whole protein sequences, protein peptide data are also non-numerical. A protein peptide is a composition of 20 amino acids, which are represented by 20 letters, such as A for Alanine and C for Cysteine (CBN, 1968). Most machine learning algorithms can only accept numerical data. Therefore, these non-numerical amino acids must be coded to numerical data using a feature extraction approach before the pattern discovery and analysis process starts, which is a common exercise in machine learning (Duda *et al.*, 2000; Bishop, 2006). There have been mainly three coding (feature extraction) approaches for converting amino acids to numerical values in peptides for the purpose of analyzing protein peptides. They convert each amino acid to a set of numerical values based on different concepts. Some are based on amino acid physio-chemical descriptors or structure properties (Tong *et al.*, 2008; Liang *et al.*, 2009; Xie *et al.*, 2010; Yang *et al.*, 2010; van Westen *et al.*, 2013), while others are based on computational or statistical methods (Cai and Chou, 1998; Chou, 2011).

The physio-chemical descriptors or structure-based coding approaches normally employ a set of descriptors to code amino acids and hence extract features for each peptide. In this way, each amino acid will be assigned a set of descriptors which are numerical. Given a set of N peptides $\mathcal{S} = (\mathbf{s}_1, \mathbf{s}_2, \ldots, \mathbf{s}_N)$, where each peptide is composed of R residues, $\mathbf{s}_n \in \mathcal{A}^R$. Suppose each amino acid is coded by d descriptors. The coding of each peptide is defined as follows:

$$\mathbf{x}_n = \phi(s_{n1})\phi(s_{n2})\ldots\phi(s_{nK}) \tag{6.4}$$

where $\phi(s_{nm}) \in \mathcal{R}^d$ is the coding vector of the mth residue of the nth peptide $s_n(s_{nm})$ and $\mathbf{x}_n \in \mathcal{R}^{\mathcal{D}}$. Note that $\mathcal{D} = R \times d$. This coding approach transfers a non-numerical peptide space to a $\mathcal{D}$-dimensional numerical space. Analysis of peptides is thus carried out in this $\mathcal{D}$-dimensional numerical space using machine learning algorithms. A model based on such a numerical feature space is defined as follows:

$$\hat{y}_n = f(\mathbf{x}_n, \mathbf{w}) \tag{6.5}$$

where $\mathbf{w}$ is a set of model parameters which are required to be estimated, $f(\cdot)$ is a function to be constructed using a machine learning algorithm, and $\hat{y}_n$ is the property prediction for the nth peptide s_n.

Table 6.1 shows the performance evaluation of the Fisher discriminant analysis (FDA) (Fisher, 1936) models and regression models constructed for six peptide datasets whose peptides were coded using the physio-chemical descriptors of Xie *et al.* (2010) for 20 amino acids. In this table, AUC stands for the area under an ROC curve (Hanley and McNeil, 1982). MCC in Table 6.1 stands for

Table 6.1. The performance evaluation (AUC and MCC) of the FDA and regression models constructed for six peptide datasets, where each peptide was coded using the physio-chemical descriptors of Xie *et al.* (2010).

Data name	FDA AUC	FDA MCC	Regression AUC	Regression MCC	References
O-linkage	0.75	0.44	0.76	0.45	Yang and Chou (2004b)
Factor Xa	0.93	0.66	0.93	0.54	Yang *et al.* (2003), Yang *et al.* (2006)
Caspase	0.90	0.66	0.93	0.65	Yang (2005b)
SARS	0.94	0.76	0.93	0.73	Yang (2005a)
HCV	0.96	0.72	0.96	0.45	Narayanan *et al.* (2002), Yang (2006)
HIV	0.88	0.64	0.88	0.60	Cai and Chou (1998), Yang *et al.* (2004), Yang and Thomson (2005)

the Matthews correlation coefficient (Matthews, 1975), which is also
referred to as the φ coefficient:

$$\varphi = \frac{\text{TN} \times \text{TP} - \text{FN} \times \text{FP}}{\sqrt{(\text{TN} + \text{FN})(\text{TP} + \text{FP})(\text{TN} + \text{FP})(\text{TP} + \text{FN})}} \qquad (6.6)$$

where TN, TP, FN, and FP are the true negative, the true positive,
the false negative, and the false positive, respectively. The true
positive stands for the number of correctly predicted cleaved (or
post-translational modified) peptides. The true negative stands for
the number of correctly predicted non-cleaved peptides. The false
positive stands for the number of non-cleaved peptides which are
falsely predicted as cleaved ones. The false negative stands for
the number of cleaved peptides which are falsely predicted as
non-cleaved. Figure 6.8 shows the ROC curves of these FDA models.

There are many amino acid descriptors available. Most of them
were developed based on protein structures. The coding approaches
using physio-chemical descriptors background or structure property
background is more biology sound because the biological properties
of amino acids are explored for coding. The pseudo-coding approach
(Chou, 2011) and the orthogonal coding approach (Cai and Chou,

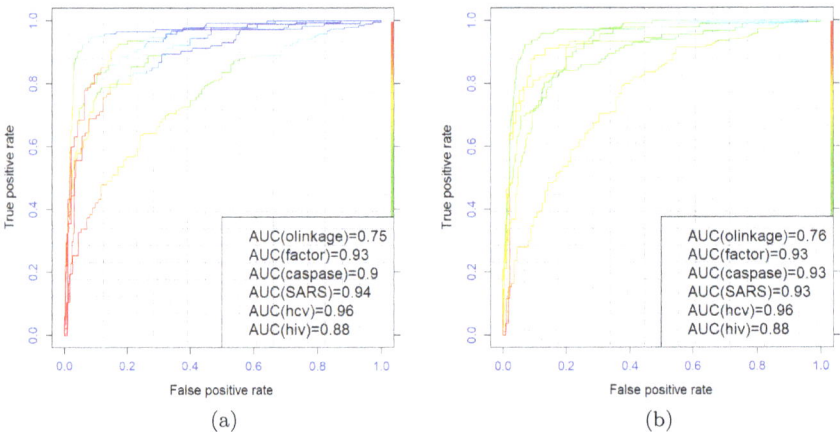

Figure 6.8.   The ROC curves of the FDA and regression models constructed for
six peptide datasets which were coded using the amino acid descriptors of Xie
et al. (2010). (a) FDA models. (b) Regression models.

1998) are, on the other hand, more computationally or statistically oriented. Using the orthogonal coding approach, each amino acid is coded using a 20-bit-long binary vector, in which only one bit is assigned a value one, meaning the presence of a specific amino acid at the residue of a peptide. The rest of the 19 bits are assigned zeros, meaning the absence of the rest of the 19 amino acids at the same residue of the same peptide. This coding approach, no doubt, has generated the space of a very large dimension. Importantly, due to the data size, it is inevitable that the approach creates some variables, which are constant (all zeros) across the whole dataset. This limitation will be discussed later in this section for the real data simulations.

Table 6.2 shows the statistics of the performance evaluation of the FDA and regression models constructed for six peptide datasets, which are coded using the orthogonal coding approach. The fourth column indicates the number of variables, which are constants (all zeros) and have to be removed from modeling and analysis to build an FDA model. These numbers show that the orthogonal coding approach may introduce a large percentage of meaningless variables.

Figure 6.9 shows the ROC curves of these models. Combining Table 6.2 and Figure 6.9, it can be seen that the orthogonal coding approach may not compete with the physio-chemical descriptor coding approach. This may be because the descriptor coding approach has at least explored some biological content of the amino acids.

Table 6.2. The performance evaluation (AUC and MCC) of the FDA and regression models constructed for six peptide datasets which are coded using the orthogonal coding approach.

Data name	FDA			Regression	
	AUC	MCC	Redundant variables	AUC	MCC
O-linkage	0.80	0.50	27	0.82	0.53
Factor Xa	0.95	0.80	86	0.95	0.62
Caspase	0.89	0.68	14	0.94	0.91
SARS	0.73	0.52	37	0.97	0.84
HCV	0.99	0.87	9	0.98	0.72
HIV	0.96	0.76	29	0.95	0.72

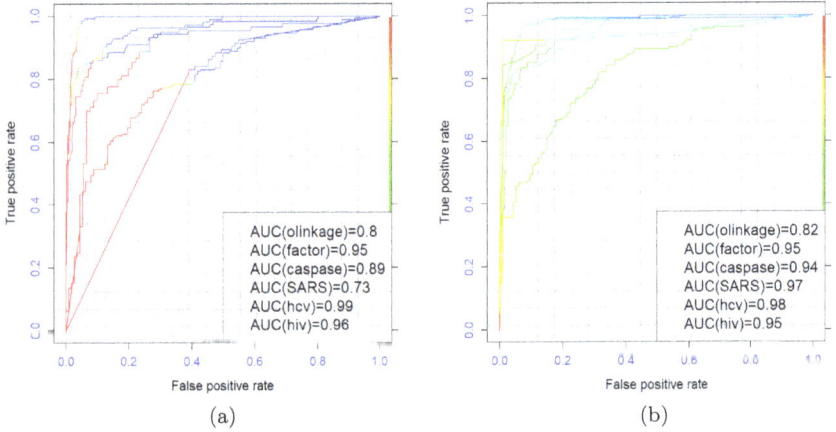

Figure 6.9. The ROC curves of the FDA and regression models constructed for six peptide datasets which were coded using the orthogonal coding approach. (a) FDA models. (b) Regression models.

6.4   Alignment between peptides

Based on the abovementioned discussion, it can be seen that cleaved peptides and post-translational modified peptides have some conserved patterns. Bearing this knowledge in mind, we now need to understand what the outcome will be if non-gapped pairwise alignment is applied to these peptides.

The non-gap pairwise alignment approach was used to align each pair of 190 SARS peptides (Yang, 2005a). During alignments, the Dayhoff mutation matrix (Dayhoff and Schwartz, 1978) was employed to measure the similarity between amino acids from each pair of peptides. Suppose $\mathcal{A}$ is a set of amino acids and two peptides with the same length are denoted by $\mathbf{s}_n \in \mathcal{A}^R$ and $\mathbf{s}_m \in \mathcal{A}^R$. The non-gap pairwise alignment score is defined as follows, where $s_{nr} \in \mathcal{A}$ and $s_{mr} \in \mathcal{A}$. Note that $\mathcal{D}(s_{nr}, s_{mr})$ is a Dayhoff mutation matrix score between amino acids s_{nr} and s_{mr}:

$$\sigma_{nm} = \sum_{r=1}^{R} \mathcal{D}(s_{nr}, s_{mr}) \tag{6.7}$$

This estimation approach for the similarity of the alignment of sequences has been widely used in both pairwise and multiple-sequence alignment algorithms. The collected alignment scores for the SARS protease cleavage peptides were organized as a matrix. Figure 6.10 shows the heatmap of this matrix. It shows that most cleaved SARS peptides have been clustered together. This is consistent with the earlier analysis that cleaved protease peptides will show great similarities.

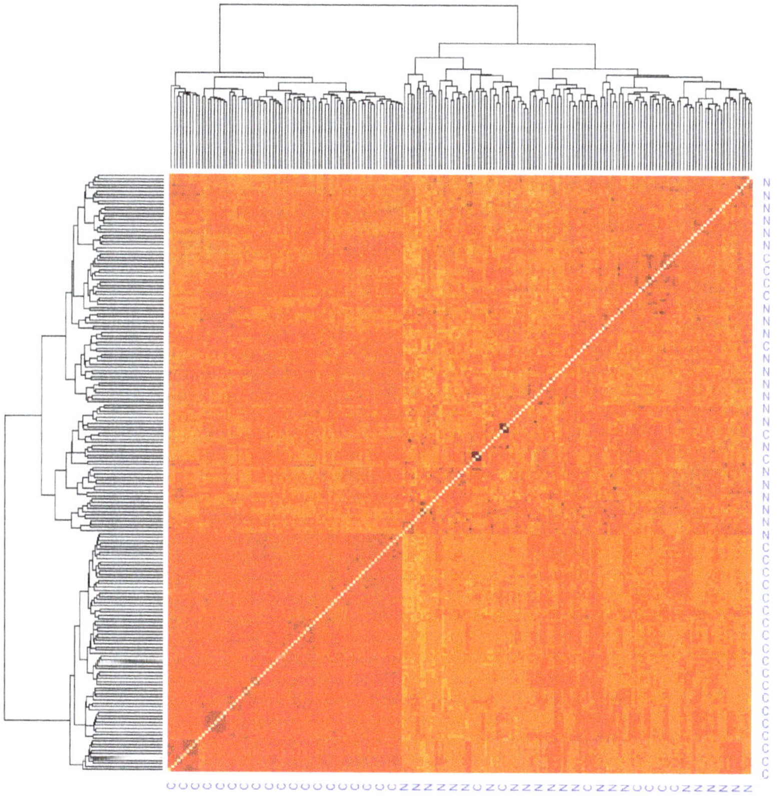

Figure 6.10. The heatmap of non-gap pairwise alignment scores for the 190 SARS peptides (Yang, 2005a). The Dayhoff mutation matrix was employed for all the alignments. N stands for non-cleaved SARS peptides and C stands for the cleaved SARS peptides.

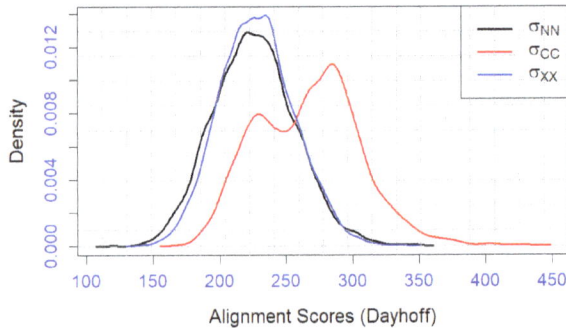

Figure 6.11. The densities of the non-gap pairwise alignment scores for the 190 SARS peptides (Yang, 2005a). The Dayhoff mutation matrix was used. σ_{CC}, σ_{NN}, and σ_{xx} stand for the score densities of the alignments between cleaved peptides, the alignments between non-cleaved peptides, and the alignments between cleaved and non-cleaved peptides, respectively.

The density of the alignment scores can also be analyzed to examine the deviation between different groups of alignment scores. The densities of three non-gap pairwise alignment scores were estimated. They are the score density of the alignments between cleaved peptides (σ_{CC}), the score density of the alignments between non-cleaved peptides (σ_{NN}), and the score density of the alignments between cleaved and non-cleaved peptides (σ_{xx}). Figure 6.11 shows the result of this density analysis for the SARS protease cleavage dataset, where the non-gap pairwise alignment scores were calculated using the Dayhoff mutation matrix. It can be seen that the score density of the alignments between cleaved peptides is on the right side. The score density of the alignments between non-cleaved peptides and the score density of the alignments between cleaved and non-cleaved peptides are on the left. In this case, σ_{CC} is the greatest. This indicates that the cleaved peptides have the largest similarity on average.

Figure 6.12 shows the heatmap of the non-gap pairwise alignment scores between the SARS cleavage peptides using the BLOSUM62 mutation matrix (Altschul *et al.*, 1990), where each score is calculated using the definition mentioned earlier, where $\mathcal{B}(s_{nr}, s_{mr})$

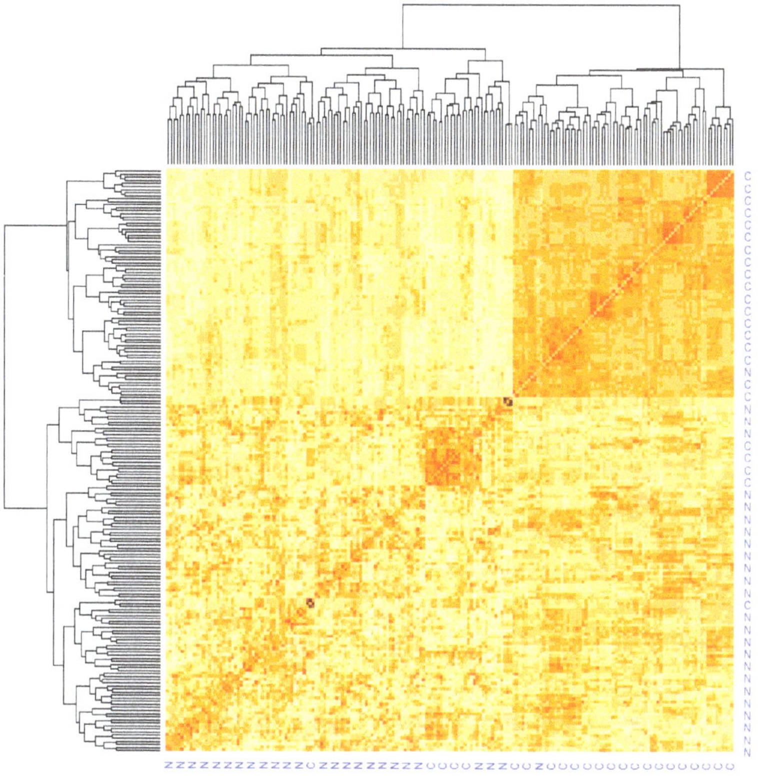

Figure 6.12. The heatmap of the non-gap pairwise alignment scores for the 190 SARS peptides. The BLOSUM62 mutation matrix was employed for the alignments. N stands for non-cleaved SARS peptides and C stands for the cleaved SARS peptides.

is a BLOSUM62 mutation matrix score between amino acids s_{nr} and s_{mr}:

$$\sigma_{nm} = \sum_{r=1}^{R} \mathcal{B}(s_{nr}, s_{mr}) \tag{6.8}$$

In the heatmap, the same pattern can be seen as well, i.e., most cleaved peptides are clustered together and most non-cleaved peptides are also clustered together. Figure 6.13 shows the density analysis for the alignment scores between different groups of the 190 SARS

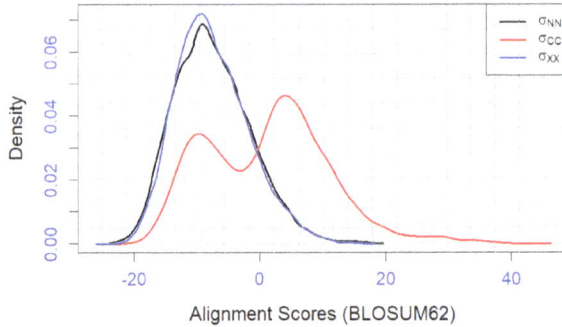

Figure 6.13. Three density functions of the non-gap pairwise alignments between the 190 SARS peptides. The BLOSUM62 mutation matrix was used. σ_{CC} stands for the score density of the alignment between cleaved peptides. σ_{NN} stands for the score density of the alignment between non-cleaved peptides. σ_{xx} stands for the score density of the alignment between cleaved and non-cleaved peptides.

cleavage peptides. The same pattern emerges. The score density of the alignments between cleaved peptides is pushed toward the right side. The score density of the alignments between non-cleaved peptides and the score density of the alignments between cleaved and non-cleaved peptides are pushed to the left side.

6.5 The properties of alignment matrix

As mentioned earlier, a cleaved peptide or a post-translational modified peptide is referred to as a functional peptide and a non-cleaved peptide, or a non-post-translational modified peptide is referred to as a non-functional peptide. As seen in the previous section, both heatmaps of the pairwise alignment scores have demonstrated a pattern where different classes of alignment scores are visualized in the map distinctly. We need to further investigate other properties of the alignment matrix. An alignment matrix is defined as follows:

$$\Sigma = \begin{Bmatrix} \sigma_{11} & \sigma_{12} & \cdots & \sigma_{1N} \\ \sigma_{21} & \sigma_{22} & \cdots & \sigma_{2N} \\ \vdots & \vdots & \ddots & \vdots \\ \sigma_{N1} & \sigma_{N2} & \cdots & \sigma_{NN} \end{Bmatrix} \tag{6.9}$$

Note that each entry (σ_{nm}) of Σ stands for the similarity between two peptides, $\mathbf{s}_n \in \mathcal{A}^R$ and $\mathbf{s}_m \in \mathcal{A}^R$. Although there is no explicit numerical approach to represent these non-numerical data, we can assume that there is an implicit function by which each peptide can be mapped to a $\mathcal{D}$-dimensional numerical feature space by the following definition:

$$\mathbf{x}_n = \phi(\mathbf{s}_n) \tag{6.10}$$

where $\mathbf{x}_n \in \mathcal{R}^\mathcal{D}$. By using the kernel trick discussed in the previous chapter, we can define a kernel space using the following transfer function:

$$\mathcal{K}(\mathbf{s}_n, \mathbf{s}_m) = \mathcal{K}(\mathbf{x}_n, \mathbf{x}_m) = \mathcal{K}(\phi(\mathbf{s}_n), \phi(\mathbf{s}_m)) \tag{6.11}$$

Although $\phi(\cdot)$ is implicit, we can still follow the discussion of the string kernel in the previous chapter to have the following relationship:

$$\mathcal{K}(\mathbf{s}_n, \mathbf{s}_m) = \mathcal{K}(\mathbf{x}_n, \mathbf{x}_m) = \mathcal{M}(\mathbf{s}_n, \mathbf{s}_m) \tag{6.12}$$

In Eq. (6.12), $\mathcal{M}$ is either the Dayhoff mutation matrix $\mathcal{D}$ or the BLOSUM62 mutation matrix $\mathcal{B}$ or the other mutation matrix. Using this approach, an alignment matrix or a kernel space $\mathbf{K}$ can be generated for a peptide datasetas shown in the following:

$$\mathbf{K} = \begin{pmatrix} \mathcal{M}(\mathbf{s}_1, \mathbf{s}_1) & \mathcal{M}(\mathbf{s}_1, \mathbf{s}_2) & \cdots & \mathcal{M}(\mathbf{s}_1, \mathbf{s}_N) \\ \mathcal{M}(\mathbf{s}_2, \mathbf{s}_1) & \mathcal{M}(\mathbf{s}_2, \mathbf{s}_2) & \cdots & \mathcal{M}(\mathbf{s}_2, \mathbf{s}_N) \\ \vdots & \vdots & \ddots & \vdots \\ \mathcal{M}(\mathbf{s}_N, \mathbf{s}_1) & \mathcal{M}(\mathbf{s}_N, \mathbf{s}_2) & \cdots & \mathcal{M}(\mathbf{s}_N, \mathbf{s}_N) \end{pmatrix} \tag{6.13}$$

Figure 6.14 shows the contour analysis of the Dayhoff mutation $\mathcal{D}$. It can be seen that all entries of $\mathcal{D}$ are positive, meaning that the value of a diagonal entry is greater than any other entries in that row or column.

We now examine whether the kernel space $\mathbf{K}$ generated based on the Dayhoff mutation matrix satisfies Mercer's condition, i.e.,

$$|\mathbf{K}| \geq 0 \tag{6.14}$$

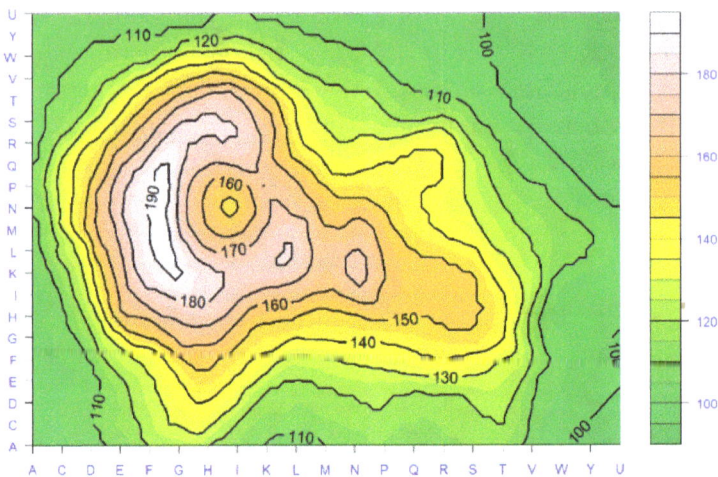

Figure 6.14. The contour of the Dayhoff matrix $\mathcal{D}$.

The **K** matrix is normalized by the following approach:

$$\mathbf{K} = \frac{\mathbf{K}}{\max(\mathbf{K})} \qquad (6.15)$$

Because of the computational power limit in R, the simulation was run based on random samples drawn from each dataset. The random sample size was 50. The process was repeated 10 times. The final determinant was the mean estimation, where $Q = 10$:

$$\sigma_{nm} = \frac{1}{Q} \sum_{r=1}^{Q} |\mathbf{K}_r| \qquad (6.16)$$

Table 6.3 shows a simulation for six peptide datasets, where it can be seen that the determinants of all the kernel spaces (the **K** matrices) are positive and definitive.

6.6   Bio-kernel machine

The bio-kernel machine was first proposed for generating a model to predict the Trypsin protease cleavage sites (Thomson *et al.*, 2003). It was based on the least-squares regression approach and is therefore

Table 6.3. The statistics of the determinants of the Gram matrices generated based on the Dayhoff mutation matrix for six peptide datasets.

Data name	Mean	Std	References
O-linkage	1	2.76E-10	Yang and Chou (2004b)
Factor Xa	1	1.25E-10	Yang *et al.* (2003), Yang *et al.* (2006)
Caspase	1	7.43E-10	Yang (2005b)
SARS	1	1.44E-10	Yang (2005a)
HCV	1	3.71E-09	Narayanan *et al.* (2002),Yang (2006)
HIV	1	5.73E-10	Cai and Chou (1998), Yang *et al.* (2004), Yang and Thomson (2005)

referred to as the LS bio-kernel machine in this book. Suppose $\mathbf{S}^{tr}$ represents a set of N training peptides, $\mathcal{H}$ represents a set of H hypothetical kernel peptides, $\mathbf{K} = (\hat{k}_{nh})_{n,h}^{N,H} = \mathcal{K}(\mathbf{S}^{tr}, \mathcal{H})$ represents the kernel space or the Gram matrix, $\mathbf{y} \in \{0, 1\}$ represents the label vector of all training peptides and $\mathbf{w} \in \mathcal{R}^H$ represents the model parameters. The LS bio-kernel machine is defined as follows, where $\mathbf{e}$ is the error vector and $\mathbf{K} \in \mathcal{R}^{N \times H}$ is a kernel matrix:

$$\mathbf{y} = \mathbf{K}\mathbf{w} + \mathbf{e} \qquad (6.17)$$

It is common to assume that the error follows a Gaussian distribution. Therefore, the likelihood function of such a model is defined as follows, where $\Sigma = \mathrm{diag}(\sigma_n^2)$:

$$\mathcal{L} = \frac{1}{(2\pi)^{N/2}\sqrt{|\Sigma|}} \exp\left(-\frac{\mathbf{e}^t \Sigma^{-1} \mathbf{e}}{2}\right) \qquad (6.18)$$

Suppose $\sigma_1^2 = \sigma_2^2 = \cdots = \sigma_N^2 = 1$. Applying the negative logarithm to Eq. (6.18) leads to the following equation, where C is a constant, which is irrelevant to the optimization process when this likelihood function is maximized:

$$-\log\mathcal{L} = \frac{1}{2}\sum_{n=1}^{N} e_n^2 + C = \frac{1}{2}\mathbf{e}^t\mathbf{e} + C \qquad (6.19)$$

Maximizing the likelihood function defined previously thus results in the error function shown in the following, where $\hat{\mathbf{y}} \in \mathcal{R}^N$ is the predicted peptide class for $\mathbf{y}$:

$$e = \frac{1}{2}(\mathbf{y} - \hat{\mathbf{y}})^t(\mathbf{y} - \hat{\mathbf{y}}) \tag{6.20}$$

This is also referred to as the least-squares (LS) bio-kernel machine. The maximization of the likelihood function leads to the solution of the model shown in the following:

$$\hat{\mathbf{w}} = (\mathbf{K}^t\mathbf{K})^{-1}\mathbf{K}^t\mathbf{y} \tag{6.21}$$

where $(\mathbf{K}^t\mathbf{K})^{-1}\mathbf{K}^t$ is called the pseudo inverse. This regression model can be enhanced by introducing a Lagrange term, which is also referred to as a regularization term as shown in the following, where λ is the Lagrange coefficient or the Lagrange multiplier:

$$\varepsilon = \frac{1}{2}(\mathbf{y} - \hat{\mathbf{y}})^t(\mathbf{y} - \hat{\mathbf{y}}) + \frac{1}{2}\lambda\mathbf{w}^t\mathbf{w} \tag{6.22}$$

The estimation of the $\mathbf{w}$ vector is thus revised as follows:

$$\hat{\mathbf{w}} = (\mathbf{K}^t\mathbf{K} + \lambda\mathbf{I})^{-1}\mathbf{K}^t\mathbf{y} \tag{6.23}$$

This model for prediction of the testing peptides utilizes the following rule:

$$\tilde{\mathbf{y}} = \tilde{\mathbf{K}}\hat{\mathbf{w}} + \mathbf{e} \tag{6.24}$$

where $\tilde{\mathbf{y}} \in \mathcal{R}^G$ is the predicted class label vector for the G testing peptides and $\tilde{\mathbf{K}}$ is defined as follows:

$$\tilde{\mathbf{K}} = \mathcal{K}(\mathbf{S}^{\text{te}}, \mathcal{H}) \tag{6.25}$$

Note that $\mathbf{S}^{\text{te}}$ stands for a set of testing peptides. Figure 6.15 shows a simplified LS bio-kernel model, where three peptides (SYLT, AYKT, and SYKT) serve as the hypothetical kernel peptides (or kernels), while one peptide (AYKT) serves as an input, which can be a training peptide or a testing peptide. This means that $\mathcal{H} =$ (SYLT, AYKT, SYKT) and $\mathbf{S}^{\text{tr}} =$ (AYKT) or $\mathbf{S}^{\text{te}} =$ (AYKT). The model output is numerical and used for classifying each training or testing peptide to a class label, which is either functional or

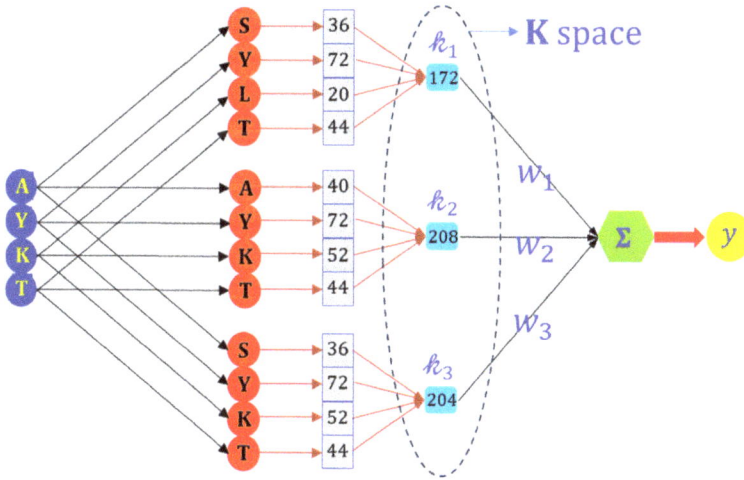

Figure 6.15. A diagram showing the working principle of LS bio-kernel, where only one output node (Σ) is used for decision-making. Three peptides are used as the hypothetical kernel peptides (or kernels) and one peptide is used as an input (training or testing peptide). The **K** space is outlined by the dashed oval. AYKT, which is in a chain of blue circles, is a training or a testing peptide. Three kernels are in chains of red circles. The nodes Σ at the right end are used for decision-making, i.e., classifying a training or a testing peptide, such as AYKT, to one of two classes, which is either functional or non-functional.

non-functional. In Figure 6.15, the node Σ has the weighted sum function from three kernels. Its definition is as follows:

$$\Sigma = \sum_{i=1}^{3} w_i k_i \qquad (6.26)$$

The value of Σ is compared with the class label for prediction performance evaluation. Table 6.4 shows the pairwise mutation scores from the Dayhoff mutation matrix. After a non-gap pairwise alignment process using the Dayhoff mutation matrix, a **K** space is formulated. It is in a three-dimensional space because of the employment of three kernels. The representative numeric feature vector in the kernel space for the input peptide (AYKT) is thus (172, 208, 204).

The model performance depends on the selection of the value of λ. Figure 6.16 shows the simulation results for the Factor Xa dataset

Table 6.4. The non-gap pairwise mutation scores for the peptides shown in Figure 6.15. The scores are calculated using the Dayhoff mutation matrix.

	A	Y	K	T
SYLT	36	72	20	44
AYKT	40	72	52	44
SYKT	36	72	52	44

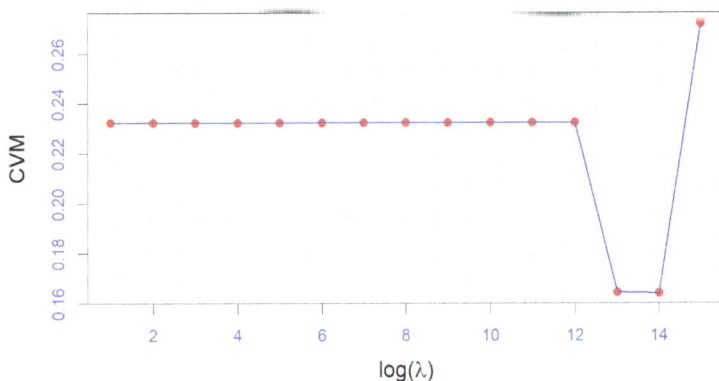

Figure 6.16. The optimization of λ value in the LS bio-kernel model for the Factor Xa dataset. CVM stands for the cross-validation measurements of the errors.

using the LS bio-kernel, where the performance was measured by the cross-validation error. From this plot, we can check when the cross-validation error reaches the minimum for a specific λ value.

The LS bio-kernel models have been constructed for six peptide datasets. The model's construction and evaluation were all based on the 5-fold cross-validation approach. By using this approach, each peptide dataset was divided into five folds. One model was constructed based on four folds of the peptides and was thus tested on the remaining other one fold of the peptides. The process was repeated five times so that each peptide is used for testing exactly once. During the construction of the model, a set of peptides from four folds of training peptides were randomly selected to serve as kernel peptides $\mathcal{H}$. The final evaluation was carried out using the

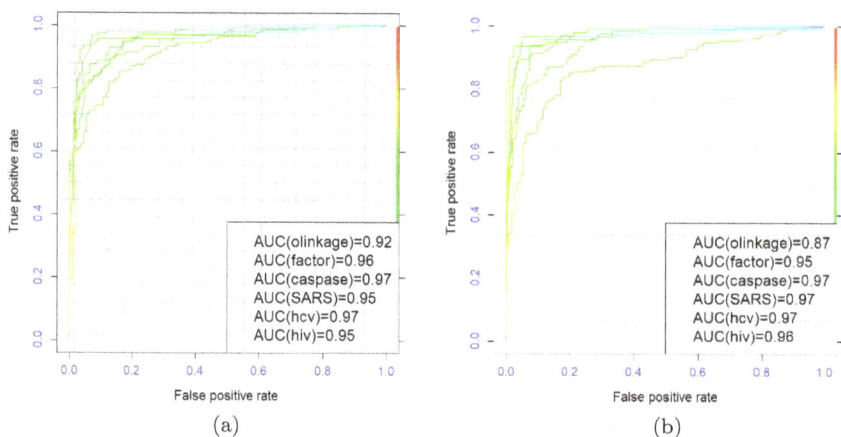

Figure 6.17. The ROC curves of the LS bio-kernel models constructed for six peptide datasets. (a) The models constructed using the Dayhoff mutation matrix. (b) The models constructed using the BLOSUM62 mutation matrix.

Table 6.5. The statistics of the measured performance (AUC and MCC) of the LS bio-kernel model constructed for six peptide datasets. Dayhoff stands for the models constructed using the Dayhoff mutation matrix and BLOSUM62 stands for the models constructed using the BLOSUM62 mutation matrix.

Data name	BLOSUM62		Dayhoff	
	AUC	MCC	AUC	MCC
O-linkage	0.92	0.66	0.87	0.60
Factor Xa	0.96	0.65	0.95	0.60
Caspase	0.97	0.86	0.97	0.88
SARS	0.95	0.80	0.97	0.84
HCV	0.97	0.65	0.97	0.64
HIV	0.95	0.70	0.96	0.69

ROC and MCC measurements. Figure 6.17 shows the ROC curves of the LS bio-kernel models constructed for six peptide datasets based on the optimal λ values. Table 6.5 shows the measured performance statistics (AUC and MCC).

When FDA is kernelized, the bio-kernel machine can also be embedded into the FDA model for modeling and analyzing peptide

datasets (Mika *et al.*, 1999; Baudat and Anouar, 2000; Li *et al.*, 2003).
It is thus referred to as the FDA bio-kernel machine. The objective
function, which needs to be maximized in a kernel space, is defined
as follows:

$$\mathcal{O} = \frac{\mathbf{w}^t \Sigma_{\mathbf{K}}^B \mathbf{w}}{\mathbf{w}^t \Sigma_{\mathbf{K}}^W \mathbf{w}} \tag{6.27}$$

where $\Sigma_{\mathbf{K}}^B$ is the between-class covariance matrix of the kernel space
$\mathbf{K}$, $\Sigma_{\mathbf{K}}^W$ is the within-class covariance matrix of the kernel space
$\mathbf{K}$, and $\mathbf{w}$ is the projection direction of model parameter vector as
mentioned earlier. The between-class covariance matrix is defined as
follows:

$$\Sigma_{\mathbf{K}}^B = (\mathbf{u}_{\mathbf{K}}^\alpha - \mathbf{u}_{\mathbf{K}}^\beta)(\mathbf{u}_{\mathbf{K}}^\alpha - \mathbf{u}_{\mathbf{K}}^\beta)^t \tag{6.28}$$

where $\mathbf{u}_{\mathbf{K}}^\alpha$ and $\mathbf{u}_{\mathbf{K}}^\beta$ are the mean vectors of the classes α and β in the
kernel space $\mathbf{K}$, respectively. Their definition is shown as follows:

$$\mathbf{u}_{\mathbf{K}}^k = \frac{1}{N_k} \sum_{n=1}^{N_k} \mathbf{k}_n \quad \forall y_n = k \tag{6.29}$$

where $\mathbf{k}_n = (\mathbf{k}_{n1}, \mathbf{k}_{n2}, \ldots, \mathbf{k}_{nH})$ and $k \in \{1, 2\}$ or $k \in \{\alpha, \beta\}$. The
within-class covariance matrix is defined as follows, where K is the
number of classes, which is two:

$$\Sigma_{\mathbf{K}}^W = \sum_{k=1}^{K} \sum_{n=1}^{N_k} (\mathbf{k}_n - \mathbf{u}_{\mathbf{K}}^k)(\mathbf{k}_n - \mathbf{u}_{\mathbf{K}}^k)^t \tag{6.30}$$

Table 6.6 shows the performance evaluation of two sets of FDA
bio-kernel machine models using AUC and MCC. One model set
was constructed using the BLOSUM62 mutation matrix and the
other model set was constructed using the Dayhoff mutation matrix.
Figure 6.18 shows the ROC curves of these models. The models work
reasonably well except for the O-linkage dataset.

The bio-kernel machine can also be incorporated with the radial
basis function model with two outputs as shown Figure 6.19, where
three peptides (SYLT, AYKT, and SYKT) serve as the kernels, while

Table 6.6. The performance (AUC and MCC) of the FDA bio-kernel models constructed for six peptide datasets using two mutation matrices. Dayhoff stands for the models constructed using the Dayhoff mutation matrix and BLOSUM62 stands for the models constructed using the BLOSUM62 mutation matrix.

	BLOSUM62		Dayhoff	
Data name	AUC	MCC	AUC	MCC
O-linkage	0.77	0.50	0.81	0.53
Factor Xa	0.94	0.74	0.94	0.68
Caspase	0.93	0.84	0.93	0.79
SARS	0.93	0.79	0.97	0.82
HCV	0.98	0.80	0.98	0.79
HIV	0.93	0.70	0.91	0.70

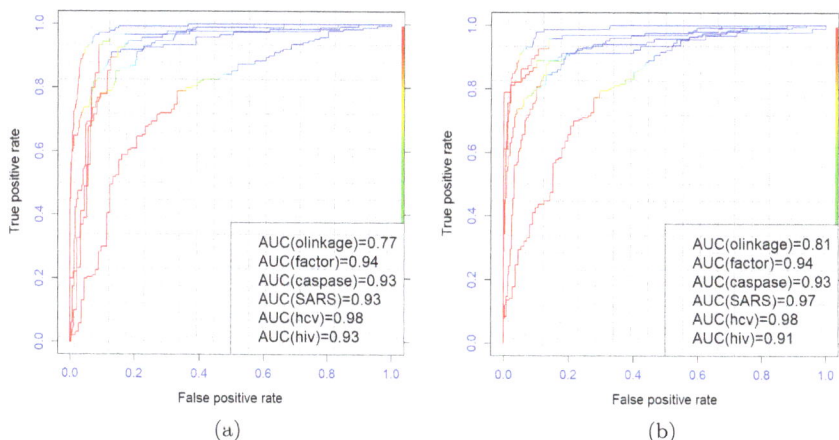

Figure 6.18. The ROC curves of the FDA bio-kernel models constructed for six peptide datasets. (a) The models constructed using the BLOSUM62 mutation matrix. (b) The models constructed using the Dayhoff mutation matrix.

one peptide (AYKT) serves as an input. Σ_α stands for the output of the non-functional peptide and Σ_β stands for the output of the functional peptide. In this model, two nodes are used for decision-making. Therefore, the model parameters are contained in a matrix

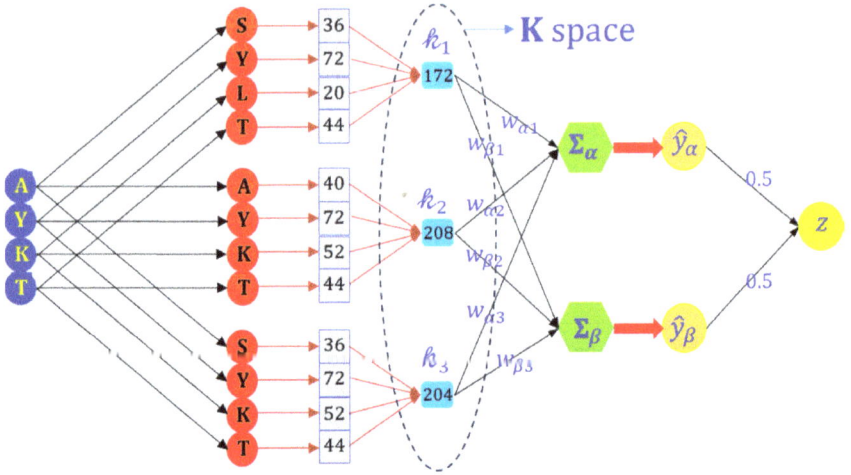

Figure 6.19. A diagram showing the working principle of a bio-kernel machine model, in which the output is transformed using the radial basis function.

$\mathbf{W}$ as shown in the following:

$$\mathbf{W} = \begin{pmatrix} w_{\alpha 1} & w_{\beta 1} \\ w_{\alpha 2} & w_{\beta 2} \\ w_{\alpha 3} & w_{\beta 3} \end{pmatrix} \qquad (6.31)$$

The output matrix is shown in the following, if there are N peptides for modeling:

$$\hat{\mathbf{Y}} = \begin{pmatrix} \hat{y}_1^{\alpha} & \hat{y}_1^{\beta} \\ \hat{y}_2^{\alpha} & \hat{y}_2^{\beta} \\ \vdots & \vdots \\ \hat{y}_N^{\alpha} & \hat{y}_N^{\beta} \end{pmatrix} \qquad (6.32)$$

Note that the outputs are transformed using the sigmoid function ρ shown as follows:

$$\begin{aligned} \hat{y}_n^{\alpha} &= \rho(\pmb{k}_n^t \mathbf{w}_{\alpha}) \\ \hat{y}_n^{\beta} &= \rho(\pmb{k}_n^t \mathbf{w}_{\beta}) \end{aligned} \qquad (6.33)$$

Because two decision-making nodes are used for classifying each testing peptide, the decision is made by comparing the magnitudes of the outputs of the two decision-making nodes. The process employs the following equation, where $\hat{y}_n$ is the prediction for the peptide $\mathbf{s}_n$:

$$\mathbf{s}_n = \begin{cases} \text{Non-functional} & \text{if } \hat{y}_n^\alpha > \hat{y}_n^\beta \\ \text{functional} & \text{if } \hat{y}_n^\alpha \leq \hat{y}_n^\beta \end{cases} \tag{6.34}$$

6.7 Orthogonal coding versus the mutation matrix

It is useful to analyze the similarity and difference between the orthogonal coding approach and the use of the mutation matrix when we model peptide data. The major difference is the working space of a model. The orthogonal coding approach constructs a model in an explicit feature space as shown in the following:

$$f := \phi^{\mathbb{O}}(\mathbf{s}_n^{\text{tr}}) \rightarrow y_n \tag{6.35}$$

where $\mathbf{s}$ is a peptide, $\phi^{\mathbb{O}}(\mathbf{s})$ is a map of $\mathbf{s}$ in the feature space, y is the corresponding class label of $\mathbf{s}$, f is a machine learning model, and $\mathbf{s}_n^{\text{tr}}$ is the nth training peptide. The use of a mutation matrix in the bio-kernel machine employs a different modeling strategy. Note that $\phi^{im}(\mathbf{s}_n^{\text{tr}})$ maps $\mathbf{s}_n^{\text{tr}}$ to a numerical feature in an implicit feature space $\mathbf{\Phi}^{im}$, while $\phi^{im}(\mathbf{s}_h^{\mathcal{H}})$ maps the hth hypothetical kernel peptide $\mathbf{s}_h^{\mathcal{H}}$ to another numerical feature in the same implicit feature space $\mathbf{\Phi}^{im}$. A collection of all the maps of $\mathbf{s}_n^{\text{tr}}$ in a kernel space is shown as follows:

$$\mathcal{K}^{\mathcal{D}}(\phi^{im}(\mathbf{s}_n^{\text{tr}}), \mathbf{\Phi}^{im}(\mathcal{H})) \tag{6.36}$$

Note that the outcome of Eq. (6.36) is a mapping point of the nth peptide in the H-dimensional kernel space $\mathbf{K}^{\mathcal{D}}$ for the nth peptide $\mathbf{s}_n^{\text{tr}}$. This means that Eq. (6.36) results in a vector of H dimension. $\mathbf{K}^{\mathcal{D}}$ stands for the kernel space and is generated using the Dayhoff mutation matrix. An H-dimensional kernel space with N data points $(\mathbf{K}^{\mathcal{D}})$ is thus defined as follows:

$$\mathcal{K}^{\mathcal{D}}(\mathbf{\Phi}^{im}(\mathbf{S}^{\text{tr}}), \mathbf{\Phi}^{im}(\mathcal{H})) \tag{6.37}$$

The bio-kernel machine aims to model the following relationship, where f^D is a machine learning model:

$$f^{\mathcal{D}} := \mathcal{K}^{\mathcal{D}}(\boldsymbol{\Phi}^{im}(\mathbf{S}^{tr}), \boldsymbol{\Phi}^{im}(\mathcal{H})) \rightarrow \mathbf{y} \qquad (6.38)$$

However, if we manipulate the data as shown in the following, we can find out some similarities and differences between the kernel space generated from orthogonally coded vectors and the kernel space generated using a mutation matrix. Rather than working in the feature space generated by the orthogonal coding approach, we can adopt a kernel approach by which an orthogonally coded space is transferred to an orthogonal kernel space. This means that the explicit coding space made by the orthogonal coding approach is not in use. Instead, it serves as an intermediate space between a peptide space and an orthogonal kernel space. A machine learning model will not be built in the orthogonally coded feature space ($\boldsymbol{\Phi}^{\mathcal{O}}(\mathbf{S}^{tr})$). Instead, a machine learning model is constructed in the orthogonal kernel space ($\mathbf{K}^{\mathcal{O}}$) converted from the orthogonally coded feature space:

$$f^{\mathcal{O}} := \mathcal{K}^{\mathcal{O}}(\boldsymbol{\Phi}^{\mathcal{O}}(\mathbf{S}^{tr}), \boldsymbol{\Phi}^{\mathcal{O}}(\mathcal{H})) \rightarrow \mathbf{y} \qquad (6.39)$$

Using this kernel function, the dot product of the orthogonally coded vectors for two different amino acids is always zero and the dot product of the orthogonally coded vectors for two identical amino acids is always one. This is the exact edit distance employed by the Sellers algorithm or the Needleman–Wunsch algorithm. For instance, the dot product of the orthogonally coded vectors for A (Alanine) and S (Serine) is zero because A is expressed by a binary string 1000000000, 00000000000 and S is expressed by another binary string 0000000000, 0000010000.

Figure 6.20 shows such a network. In this network, the comparison between the testing peptide AYKT and the kernel peptide SYLT results in a score of two, using the dot product of the orthogonally coded vectors for the peptides. Two pairs of the residues in two peptides are identical: Y and T. The mismatched pairs are A − S and K − L. The dot product of the orthogonally coded vectors between

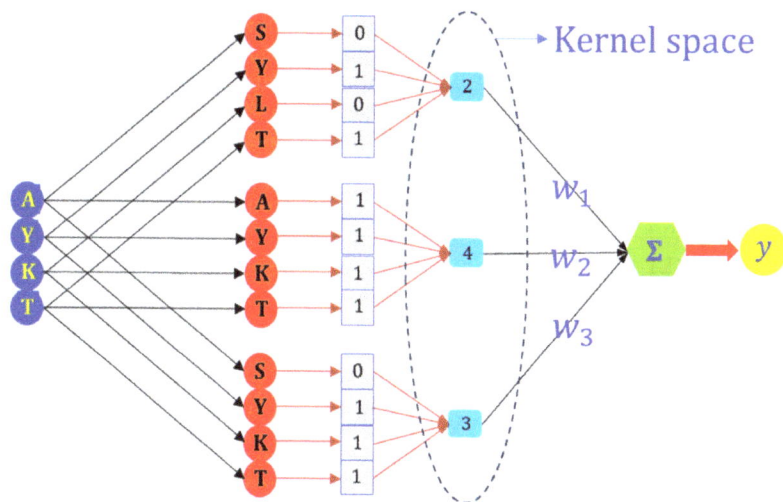

Figure 6.20. The diagram of a kernel model using the dot product of the orthogonally coded vectors for a peptide data space.

the testing peptide and the second kernel peptide (AYKT) is four because these two 4-residue peptides are identical. Because there is only one residue difference (mismatch) between the testing peptide and the third kernel peptide (SYKT), the dot product between their orthogonally coded vectors is thus three. It can be seen that f^O generates an edit distance kernel space, while f^D generates a non-edit distance kernel space. $\mathbf{K}^{\mathbb{O}}$ can be considered as a naïve version of the kernel space $\mathbf{K}^{\mathcal{D}}$. Moreover, $\mathbf{K}^{\mathcal{D}}$ is biologically sound compared to $\mathbf{K}^{\mathbb{O}}$.

Table 6.7 shows a simple comparison in terms of the dynamics of the similarity (distance) measurements between a testing peptide and three kernel peptides as illustrated in Figure 6.20. It can be seen that the change rates of the alignment scores between the testing peptide and three kernel peptides calculated using the Dayhoff mutation matrix are very different from those calculated using the dot products of the orthogonally coded vectors for the peptides. For instance, using the Dayhoff mutation matrix for peptide alignments, the alignment score change rate is 20.9% from the alignment between AYKT and

Table 6.7. The comparison of the dynamics of the peptide similarity (distance) measurements between the bio-kernel approach using the Dayhoff mutation matrix and the dot products of the orthogonally coded vectors for the peptides shown in Figure 6.20.

AYKT	Dayhoff	Increase %	Orthogonal	Increase %
SYLT	172		2	
SYKT	204	18.6%	3	50%
AYKT	208	20.9%	4	100%

SYLT to the alignment between AYKT and AYKT. However, using the dot product of the orthogonally coded vectors, the score change rate is 100% from the pair of AYKT and SYLT to the pair of AYKT and AYKT.

In fact, the Dayhoff mutation matrix was developed based on the observation of the amino acid mutations across evolutionary generations. Different pairs of amino acids have different mutation probabilities across evolutionary generation (Dayhoff and Schwartz, 1978). For instance, the mutation scores of A (Alanine), Y (Tyrosine), K (Lysine), and T (Threonine) to S (Serine), Y, L (Leucine), and T are 36, 72, 20, and 44, respectively, in this example. However, this difference cannot be reflected by the orthogonal coding approach.

Further analysis can be carried out by reviewing the Sellers algorithm (Sellers, 1974) which was developed for whole-sequence alignments. The algorithm employs a metric to measure the dissimilarity (distance) between two sequences, which is referred to as the edit distance. Table 6.8 shows the metric used by the Sellers algorithm. It can be seen that the metric is very similar to the dot product result of the orthogonally coded data.

The Sellers algorithm and the Needleman–Wunsch algorithm (Needleman and Wunsch, 1970), as well as the Smith–Waterman algorithm (Smith and Waterman, 1981), have the limitation of reflecting the true relationship between sequences during sequence alignmentas mentioned in the previous chapter. After the development of the Dayhoff mutation matrix (Dayhoffand Schwartz, 1978),

Table 6.8. The metric employed by the Sellers algorithm for aligning sequences. x and y are two amino acids. Adapted from Yang (2022a).

$d(x, y) = 0$	If $x = y$, match, no distance
$d(x, y) = 1$	If $x \neq y$, mismatch
$d(x, -) = 1$	Gap inserted into the 2nd sequence
$d(-, y) = 1$	Gap inserted into the 1st sequence

more biologically sound sequence alignment algorithms were developed based on various mutation matrices for sequence alignment. These new sequence alignment algorithms generated more reliable and accurate sequence alignment results and hence more meaningful protein function annotations (Lipman *et al.*, 1989).

We now use the O-linkage data (Yang and Chou, 2004b) for a further comparison between the kernel space $\mathbf{K}^{\mathbb{O}}$ generated using the dot products of the orthogonally coded data and the kernel space $\mathbf{K}^{\mathbb{D}}$ generated using the Dayhoff mutation matrix. Two cleaved peptides (EPLVSTSEP and TTTSSSVSK) were randomly selected as the hypothetical kernel peptides from all peptides to act as the two coordinates for better visualization. This means that both $\mathbf{K}^{\mathbb{O}}$ and $\mathbf{K}^{\mathbb{D}}$ are in a two-dimensional space. All peptides were then mapped to these two-dimensional spaces coordinated by these two kernel peptides. Figure 6.21 shows these three kernel spaces generated for the O-linkage peptide data, where we find that all show certain degrees of complexity. This is not a surprise because only the two cleaved peptides that are served as kernels will definitely not contain the complete discrimination information for analyzing the peptides. However, the kernel space $\mathbf{K}^{\mathbb{O}}$ shows that many more peptides have identical dot product values (identical mapping locations) although they have different labels (a mixture of cleaved and non-cleaved peptides). In the kernel space $\mathbf{K}^{\mathbb{D}}$, this complexity is significantly reduced.

Table 6.9 shows the comparison. In Table 6.9, $\mathbf{K}^{\mathbb{O}}$ stands for the kernel space generated using the dot products of the orthogonally coded vectors. $\mathbf{K}^{\mathbb{D}}$ stands for the bio-kernel space generated using

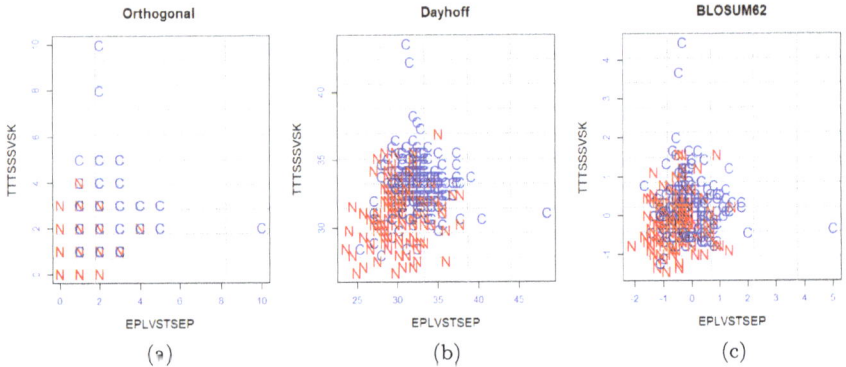

Figure 6.21. Three kernel spaces generated for the O-linkage peptide dataset (Yang and Chou, 2004b). (a) The kernel space generated from the orthogonally coded features. (b) The bio-kernel space supported by the Dayhoff mutation matrix. (c) The bio-kernel space supported by the BLOSUM62 mutation matrix. C stands for cleaved peptides and N stands for non-cleaved peptides.

Table 6.9. The numbers of the pairs of the O-linkage peptides of different classes, which were mapped to identical locations in the kernel spaces.

Kernel space	Pairs mapped at the identical locations
$\mathbf{K}^{\circleddash}$	1083
$\mathbf{K}^{\mathcal{D}}$	44
$\mathbf{K}^{\mathcal{B}}$	38

the Dayhoff mutation matrix and $\mathbf{K}^{\mathcal{B}}$ stands for the bio-kernel space generated using the BLOSUM62 mutation matrix. We can see that 1083 pairs of peptides of different classes were mapped to identical locations in the two-dimensional space $\mathbf{K}^{\circleddash}$. In contrast, only 44 and 38 pairs of peptides of different classes were mapped to identical locations in $\mathbf{K}^{\mathcal{D}}$ and $\mathbf{K}^{\mathcal{B}}$, respectively. Table 6.10 shows a part of the results of this problem, i.e., peptides of different classes were mapped to identical locations in a kernel space.

Table 6.10. The pairs of peptides of different classes which were mapped to identical locations in the kernel spaces for the O-linkage peptide data. ϑ stands for the status of a peptide, C stands for a cleaved peptide, and N stands for a non-cleaved peptide.

Spaces	Peptide one	ϑ	Location		Peptide two	ϑ	Location	
$\mathbf{K}^{\Theta}$	TPPPTSGPT	C	2	3	DSVTSTFSK	N	2	3
	TPPPTSGPT	C	2	3	TEHLSTLSE	N	2	3
$\mathbf{K}^{\mathcal{D}}$	ARSPSPSTQ	C	32	34.7	TEHLSTLSE	N	32	34.7
	ARSPSPSTQ	C	32	34.7	PTVFTRVSA	N	32	34.7
$\mathbf{K}^{\mathcal{B}}$	SSIATVPVT	C	−0.2	−0.4	PVLESFKVS	N	−0.2	−0.4
	SSIATVPVT	C	−0.2	−0.4	AQSVTLNSY	N	−0.2	−0.4

6.8 Summary

This chapter introduced the properties of peptide data. The introduction has shown that peptide pattern discovery and analysis are different from whole-sequence pattern discovery and analysis. The former mainly aims to make a predictive model to predict functional sites within shorter protein sequences or peptides. The latter mainly aims to annotate the structure or function for a whole protein through sequence alignment. The former analysis focuses on local areas of a protein, while the latter targets the global protein structure and function. Most machine learning algorithms can be directly used for peptide pattern discovery and analysis under the condition that amino acids have been well coded to numerical data. From this, three conventional coding approaches were briefly introduced. This chapter compared their differences and similarities. Afterward, this chapter introduced the least-squares regression bio-kernel machine as well as the Fisher discriminant analysis bio-kernel machine with applications to six peptide datasets. These bio-kernel machines were developed based on the employment of the mutation matrix, especially the Dayhoff mutation matrix and the BLOSUM62 mutation matrix. Similar to whole-sequence alignment, the employment of a mutation matrix in peptide pattern discovery and analysis enables a constructed model to be more biologically sound, i.e., to truly reflect the evolutionary background of protein sequences or amino acids. The applications of the least-squares regression bio-kernel machine and the Fisher discriminant analysis bio-kernel machine to six peptide datasets demonstrate better model generalization performance. In the following chapters, this bio-kernel machine will be revised into some advanced formats for better generalization capability or parsimoniousness model structures.

Chapter 7

Advanced Bio-Kernel Machines

The least-squares regression bio-kernel machine and the Fisher discriminant analysis bio-kernel machine introduced in the last chapter were generated using basic machine learning algorithms. In this chapter, advanced bio-kernel machines are introduced and demonstrated. These advanced bio-kernel machines are developed to improve model performance, including parsimonious model structures, the generalization capability, and intelligent knowledge discovery. Some of the advanced bio-kernel machines introduced in this chapter have appeared in several previous publications. They include the shrinkage models, the orthogonal models, the Bayesian models, and the intelligence models. Deep bio-kernel machines are also introduced in this chapter, which incorporate support vector machines and relevance vector machines to further improve the model's generalization capability.

7.1 Parsimonious bio-kernel machines

First, we introduce parsimonious bio-kernel machines. It is known that a kernel machine works in a kernel space which is transferred from a feature space. Suppose $\mathbf{x}_n \in \mathcal{D}$ is a peptide of an input peptide set $\mathcal{D}$ and $\mathbf{x}_h \in \mathcal{H}$ is a peptide of a hypothetical peptide set $\mathcal{H}$. The size of $\mathcal{H}$ is H. We also assume that $\phi(\mathbf{x}_n) \in \mathcal{R}^d$ and $\phi(\mathbf{x}_h) \in \mathcal{R}^d$ are two vectors in a feature space. The kernel space is defined as follows:

$$\mathbf{K} = \{\mathcal{K}(\phi(\mathbf{x}_n), \phi(\mathbf{x}_h))\}_{n,h=1}^{N,H} \tag{7.1}$$

where $\mathbf{K} \in \mathcal{R}^{N \times H}$ is a kernel space, which can have a very high dimension depending on H. It is often known that not every part of this high-dimensional space is informative or contributing to model performance. Some could be useless or some could have a negative impact. Removing these variables so as to improve model performance has been researched in the literature and the theory of this type of practice is referred to as Occam's razor (Charlesworth, 1956; Soklakov, 2002). The theory claims that a simpler model may generalize well compared with a more complicated model under the condition that the simpler model is well constructed (Duda *et al.*, 2000; Bishop, 2006; Dowe *et al.*, 2007).

Among various algorithms for generating parsimonious models, the ridge regression (Hoerl, 1962; Hoerl and Kennard, 1970a, 1970b; McDonald, 2009) and Lasso (least absolute shrinkage and selection operator) regression (Santosa and Symes, 1986; Tibshirani, 1996, 1997) are two typical ones. As shown in the previous chapter, ridge regression introduces a Lagrange regularization term. Suppose there is a peptide dataset $\mathcal{D} = (\mathbf{s}_1, \mathbf{s}_2, \ldots, \mathbf{s}_N) \in \mathcal{A}^{N \times R}$, in which each peptide $\mathbf{s}_n$ has R residues and there is an implicit feature space $\mathbf{\Phi}^{\mathcal{D}} = (\phi_1, \phi_2, \ldots, \phi_N) \in \mathcal{R}^{N \times d}$ for all the peptides. Each peptide is thus mapped to an implicit numerical d-dimensional feature space by using a feature extraction function. Note that we use $\phi_n = \phi(\mathbf{s}_n)$ for simplicity. A set of hypothetical kernel peptides is defined by $\mathcal{H} = (\mathbf{s}_1, \mathbf{s}_2, \ldots, \mathbf{s}_H) \in \mathcal{A}^{H \times R}$. This set is named as a hypothetical model because these hypothetical kernel peptides are used to represent the problem. Some of them may represent the true story but others may not. In a geometrical view, each of them stands for one coordinate in the kernel space. A more practical approach to generate a kernel space is to use a different set of peptides to represent kernel peptides as implemented in the LS bio-kernel. Therefore, we can avoid the overlap between the training peptides and the kernel peptides if

$$\mathcal{D} \cap \mathcal{H} = \emptyset \qquad (7.2)$$

where $\emptyset$ stands for an empty set. Suppose there are N training peptides and H hypothetical kernel peptides. A transformation using a kernel function maps this implicit feature space to a kernel space as shown in the following:

$$\mathbf{K} = \mathcal{K}(\mathbf{\Phi}^{\mathcal{H}}, \mathbf{\Phi}^{\mathcal{D}}) \tag{7.3}$$

A kernel space for constructing a bio-kernel machine model is implemented using the following approach:

$$\mathbf{K} = \mathcal{K}(\mathbf{\Phi}^{\mathcal{H}}, \mathbf{\Phi}^{\mathcal{D}}) = (\hbar_{nh})_{n=1,h=1}^{N,H} \tag{7.4}$$

Note that the use of $\mathcal{K}(\mathbf{\Phi}^{\mathcal{H}}, \mathbf{\Phi}^{\mathcal{D}})$ means that we measure the mutation from $\mathcal{H}$ to $\mathcal{D}$. The dimension of this kernel space is $N \times H$. The LS bio-kernel can be expressed as follows, where $\mathbf{w} \in \mathcal{R}^{H}$ is a set of model parameters, which are required to be estimated:

$$\mathbf{y} = \mathbf{Kw} + \mathbf{e} \tag{7.5}$$

Note that $\mathbf{y} \in \{0, 1\}^{N}$ is a vector of the known class label of the training peptides. To estimate $\mathbf{w}$, the ordinary least-squares error (LSE) function is employed. The LSE function is defined as follows:

$$\mathbf{e} = \frac{1}{2}(\mathbf{y} - \hat{\mathbf{y}})^{t}(\mathbf{y} - \hat{\mathbf{y}}) = \frac{1}{2}(\mathbf{y} - \mathbf{Kw})^{t}(\mathbf{y} - \mathbf{Kw}) \tag{7.6}$$

Note that $\hat{\mathbf{y}} = \mathbf{Kw} \in \mathcal{R}^{N}$ in the previous equation represents the estimated class labels corresponding to the true or correct class labels $\mathbf{y}$. The estimation of the model parameters can follow different strategies. A typical one is the ordinary LSE which minimizes $\mathbf{e}$, leading to the solution shown as follows:

$$\hat{\mathbf{w}}^{\mathcal{O}} = (\mathbf{K}^{t}\mathbf{K})^{-1}\mathbf{K}^{t}\mathbf{y} \tag{7.7}$$

where $\hat{\mathbf{w}}^{\mathcal{O}}$ stands for the ordinary LSE estimation of the model parameter.

Each model parameter $\hat{w}^{\mathcal{O}} \in \mathcal{R}$ is associated with a hypothetical kernel peptide. Using the concept of regression analysis, the significance of each kernel peptide can be evaluated. To examine whether a kernel peptide has a significant impact on the classification performance of a model, a statistical significance test can be employed. A null hypothesis is set to each peptide, $\mathcal{H}_0 : \hat{w}_h^{\mathcal{O}} = 0$, followed by the calculation of a statistic called the t-statistic (Weisberg, 2005). Because $\hat{w}_h^{\mathsf{O}}$ is the estimated regression coefficient for the hth hypothetical kernel peptide, the t-test statistic is defined as follows:

$$t_h = \frac{\hat{w}_h^{\mathsf{O}}}{\text{se}(w_h^{\mathsf{O}})} \qquad (7.8)$$

where $\text{se}(\hat{w}_h^{\mathsf{O}})$ is calculated using the following formula with $\mu_h = \sum_{n=1}^{N} \hbar_{nh}/N$:

$$\text{se}(\hat{w}_h^{\mathsf{O}}) = \sqrt{\frac{\frac{1}{N-2}(\mathbf{y} - \hat{\mathbf{y}})^t (\mathbf{y} - \hat{\mathbf{y}})}{\sum_{n=1}^{N}(\hbar_{nh} - \mu_h)^2}} \qquad (7.9)$$

Figure 7.1 shows an application of this procedure to the SARS peptide dataset (Yang, 2005a). It can be seen that some kernel peptides may not play an important role for the discrimination between cleaved peptides and non-cleaved peptides. Therefore, a parsimonious model with satisfactory generalization capability could be an alternative by selecting a subset of kernel peptides with the lowest p-values.

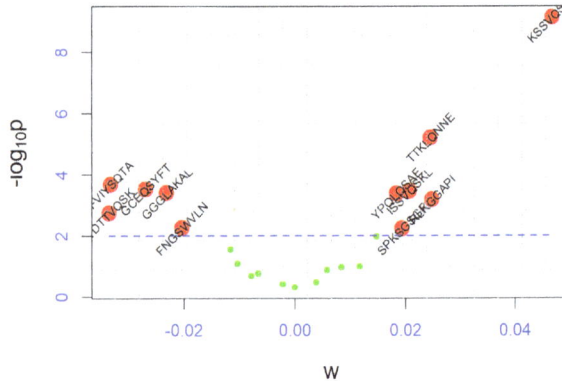

Figure 7.1. An illustration of the significance analysis of the hypothetical kernel peptides for the SARS peptide dataset. The cutting t-test p value for discriminating between insignificant and significant kernel peptides was 0.01. The horizontal dashed line stands for this cutting p value. The big dots stand for the significant kernel peptides with their p values less than 0.01. The small dots stand for the insignificant kernel peptides whose p value are greater than 0.01.

This discussion shows two facts. First, not all the bio-kernels play an equally important role in a model. Second, there is a space to generate a parsimonious model with satisfactory generalization performance if the contribution of the bio-kernels can be well evaluated. Therefore, some algorithms for generating parsimonious models are introduced in the following.

It is well known that there is a problem in linear regression models, called the ill-conditioned problem. It means that the matrix $\mathbf{K}^t\mathbf{K}$ shown in Eq. (7.7) may not be invertible, i.e., its determinant may be zero. The Tikhonov regularization was therefore proposed by introducing the Lagrange term (Hoerl, 1962; Hoerl and Kennard, 1970a, 1970b; McDonald, 2009; Kilmer *et al.*, 2007; Gazzola *et al.*, 2015), which transforms the ordinary LSE to the ridge regression as shown in the following:

$$\mathbf{e} = \frac{1}{2}(\mathbf{y} - \mathbf{K}\mathbf{w}^\lambda)^t(\mathbf{y} - \mathbf{K}\mathbf{w}^\lambda) + \frac{1}{2}\lambda((\mathbf{w}^\lambda)^t\mathbf{w}^\lambda - C) \qquad (7.10)$$

where $\mathbf{w}^\lambda$ is the model parameter which is going to be estimated using the ridge regression approach, C is a constant, and $\lambda \geq 0$ is the Lagrange coefficient or multiplier. Minimizing $\mathbf{e}$ leads to the estimation of $\mathbf{w}^\lambda$ as shown in the following:

$$\hat{\mathbf{w}}^\lambda = (\mathbf{K}^t\mathbf{K} + \lambda\mathbf{I})^{-1}\mathbf{K}^t\mathbf{y} \qquad (7.11)$$

When $\lambda = 0$, the ridge regression turns back into the ordinary LSE. Practically, the determination of the optimal value for λ is not straightforward and a grid search approach based on the cross-validation is commonly used. A better interpretation of the optimization of λ can be done using the Bayesian linear regression. The posterior probability of a Bayesian linear regression model is defined as follows (Buerkner, 2017):

$$p(\mathbf{w}^\lambda|\mathbf{y}, \mathbf{K}) = \frac{p(\mathbf{y}|\mathbf{K}, \mathbf{w}^\lambda, \Sigma) \times p(\mathbf{w}^\lambda|\alpha)}{p(\mathbf{y}|\mathbf{K})} \qquad (7.12)$$

where Σ stands for the covariance matrix and α stands for the hyper-parameter of the *a priori* structure for $\mathbf{w}^\lambda$. The likelihood function is defined as follows:

$$p(\mathbf{y}|\mathbf{K}, \mathbf{w}^\lambda, \Sigma) = \frac{1}{(2\pi)^{N/2}\sqrt{\Sigma}} \exp\left(-\frac{1}{2}(\mathbf{y} - \hat{\mathbf{y}})^t \Sigma^{-1} (\mathbf{y} - \hat{\mathbf{y}})\right)$$

(7.13)

A Gaussian *a priori* structure is defined as follows, where d stands for the number of independent variables (or data dimension):

$$p(\mathbf{w}^\lambda|\alpha) = \left(\frac{\alpha}{2\pi}\right)^{d/2} \exp\left(-\frac{\alpha}{2}(\mathbf{w}^\lambda)^t \mathbf{w}^\lambda\right) \qquad (7.14)$$

Applying the negative logarithm to Eq. (7.12) results in the following equation, where C is a constant:

$$-\log p(\mathbf{w}^\lambda|\mathbf{y}, \mathbf{K})$$
$$\propto \frac{\beta}{2}\|\mathbf{K}^t\mathbf{w}^\lambda - \mathbf{y}\|^2 + \frac{\alpha}{2}(\mathbf{w}^\lambda)^t \mathbf{w}^\lambda - \frac{N}{2}\log\beta \qquad (7.15)$$
$$- \frac{d}{2}\log\alpha + C$$

Suppose the previous equation is reorganized. The resulting model is in a format similar to the ridge regression model, where $\lambda = \alpha/\beta$:

$$\mathbf{e} = \|\mathbf{K}^t\mathbf{w}^\lambda - \mathbf{y}\|^2 + \lambda(\mathbf{w}^\lambda)^t \mathbf{w}^\lambda \qquad (7.16)$$

The Lagrange coefficient λ is used to trade off between the model fitness and the model complexity or the model parsimoniousness. As shown earlier, when $\lambda = 0$, a ridge regression model is degenerated to an ordinary regression model. However, if $\lambda \to \infty$, the model parameters are shrunk to zero ($\mathbf{w}^\infty \to \mathbf{0}$), i.e., the most parsimonious model, but completely useless.

The ridge regression is also referred to as the $\mathcal{L}_2$-regularization in contrast to the Lasso regression. The latter is also called the $\mathcal{L}_1$-regularization due to the constraint it employs (Tibshirani, 1996; Efron *et al.*, 2004). The $\mathcal{L}_1$-regularization employed in a Lasso learning process is defined as follows, where s is a positive constant:

$$\sum_{h=1}^{H} |w_h^\lambda| \le s \qquad (7.17)$$

The error or objective function of a Lasso regression model is defined as follows, where $\lambda > 0$ and H stands for the number of hypothetical kernel peptides:

$$\mathbf{e} = (\mathbf{K}^t\mathbf{w}^\lambda - \mathbf{y})^t(\mathbf{K}^t\mathbf{w}^\lambda - \mathbf{y}) + \lambda\sum_{h=1}^{H}|w_h^\lambda| \tag{7.18}$$

This $\mathcal{L}_1$ constraint has a more stringent impact on model parameters $\mathbf{w}$. Suppose a model has two variables (or kernel peptides); there are thus two model parameters, w_1 and w_2. Having applied the $\mathcal{L}_1$ constraint, it is clear that $|w_1| + |w_2| \leq s$. Therefore, the relationship between $|w_1|$ and $|w_2|$ is complementary. This means that as one of them increases, the other decreases. We can assume that both of them are positive values for simplicity. Suppose w_1 is increased by δ, i.e.,

$$w_1^{\text{new}} = w_1^{\text{old}} + \delta \tag{7.19}$$

Because $w_2 = s - w_1$, the new value of w_2 is as shown in the following:

$$w_2^{\text{new}} \leq s - (w_1 + \delta) < s - w_1 = w_2^{\text{old}} \tag{7.20}$$

The shrinkage speed of model parameters in a Lasso regression model is faster than that in a ridge regression model due to the use of the $\mathcal{L}_1$ constraint. To explain this, a model with two kernel peptides is used again. Figure 7.2 helps explain this analysis. Let us fix w_2 in an unchanged manner. Moreover, the same constraint constant is applied to both the ridge regression model and the Lasso regression model. The constant is denoted by C. The two models thus have different values for w_1:

$$\begin{aligned} w_1^{\text{Ridge}} &= \sqrt{s - w_2} = \sqrt{C} \\ w_1^{\text{Lasso}} &= s - w_2 = C \end{aligned} \tag{7.21}$$

Equation (7.21) thus leads to the following inequality:

$$w_1^{\text{Ridge}} = \sqrt{C} < C = w_1^{\text{Lasso}} \tag{7.22}$$

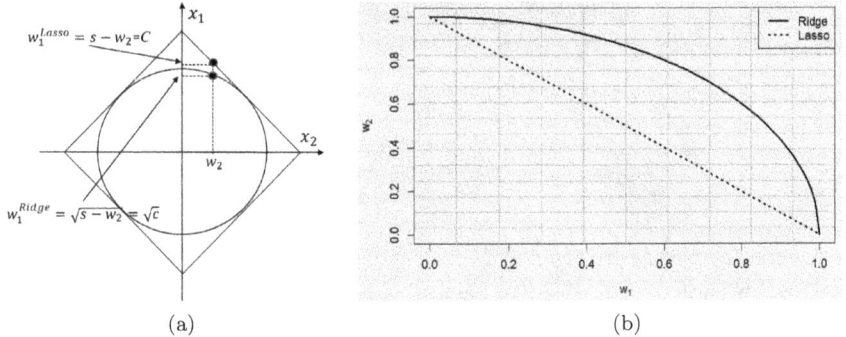

Figure 7.2. A comparison of regression coefficient shrinkage between a ridge regression model and a Lasso regression model. (a) The schematic comparison. (b) The numerical simulation comparison. Adapted from Yang (2022a).

It can be seen that the model parameters in a Lasso regression model shrink faster. This is illustrated in Figure 7.2(a). Moreover, Figure 7.2(b) provides a numeric simulation, which further proves this relationship. It can be seen that w_2 drops much faster in a Lasso regression model compared with a ridge regression model when w_1 increases by the same magnitude in both models.

It must be noted that there is no analytical solution to a Lasso regression model and the quadratic programming approach is used for estimating model parameters for a Lasso regression model (Garey and Johnson, 1979).

Figure 7.3 shows the simulation result of a ridge bio-kernel model constructed for the SARS peptide dataset (Yang, 2005a). Around 64 peptides (about 20%) were randomly selected and used as the hypothetical kernel peptides for this analysis. The values for λ were between 0.1 and 100. One ridge bio-kernel model was constructed for each λ value. The best model was that with the smallest error. After the ridge bio-kernel model for this dataset has been constructed, significance analysis was carried out. Based on the cutting p value of 0.01, ten significant kernel peptides came out as printed in the plot.

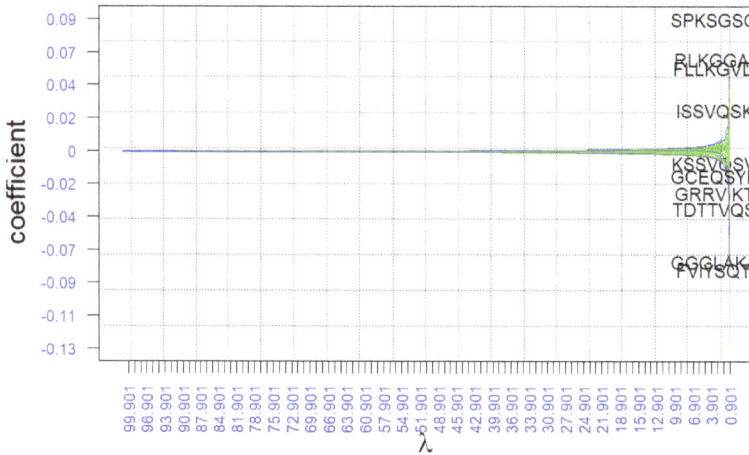

Figure 7.3. The ridge bio-kernel model constructed for the SARS peptide dataset. The cutting p value was set to 0.01, by which ten of the 64 kernel peptides were finally selected as the significant ones. The significant kernel peptides have been printed in this plot and their coefficients are represented by the thick lines. The coefficients stand for the model parameters.

Figure 7.4 further shows how the performance of the ridge bio-kernel model varies along with the number of employed kernel peptides. As expected, the most parsimonious model may not have the best generalization performance. The plot shows that the performance dropped quickly when employing too few kernel peptides. For instance, the AUC dropped to nearly 0.7 when the number of the employed kernel peptides was below five.

Figure 7.5 shows the selection of significant kernel peptides based on significance analysis for the ridge bio-kernel model constructed for the same SARS data used earlier.

Figure 7.6 shows the performance evaluation of the Lasso bio-kernel models constructed based on the different numbers of hypothetical kernel peptides. The result also shows that the best performance (AUC) may not be of the model with the smallest or most hypothetical kernel peptides.

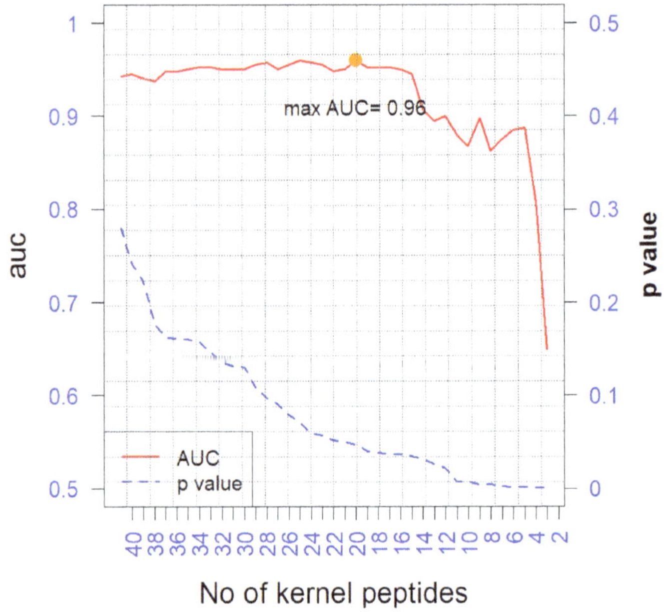

Figure 7.4. The performance evaluation (AUC) for the ridge bio-kernel model constructed for the same SARS peptide dataset shown in Figure 7.3. The number of hypothetical kernel peptides was selected from two to 40. For each selection of the kernel peptide number, the p value and AUC were measured. The best performance (maximum AUC) is pointed in the plot as a dot, where the maximum AUC was 0.96.

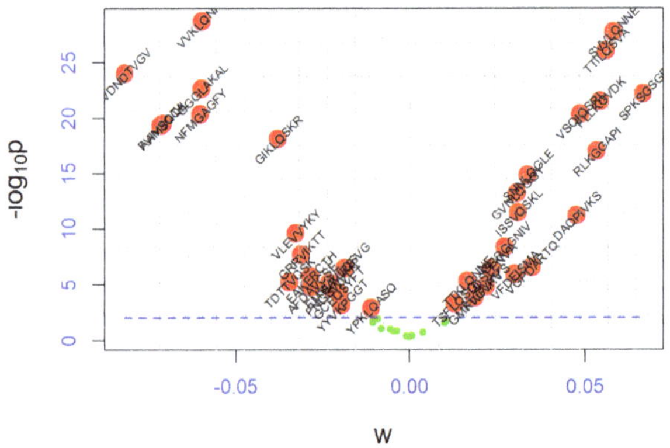

Figure 7.5. The selection of the significant kernel peptides using ridge significance analysis.

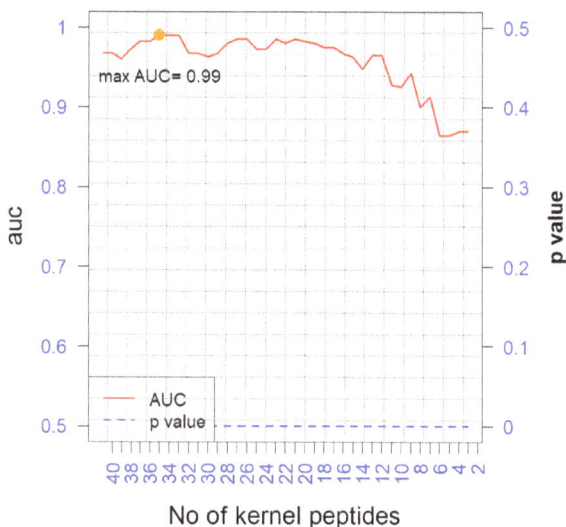

Figure 7.6. The performance evaluation of the Lasso bio-kernel model constructed for the SARS peptide data.

7.2 Orthogonal bio-kernel machine

Most experimental data are non-orthogonal, meaning that variables are mutually correlated. The non-orthogonality in experimental data may have a negative impact on the model's generalization capability. Therefore, data orthogonality has long been emphasized in statistics and machine learning (Carroll and Ruppert, 1996; Papoulis and Pillai, 2002) to deal with non-orthogonal experimental data (Gilmour and Goos, 2009; Luong *et al.*, 2022). Non-orthogonal experimental data means that variables in that dataset are mutually correlated. If two variables in such a non-orthogonal dataset are highly correlated, picking one of them and skipping the other one to analyze the data cannot help one discover the full pattern of the dataset. Having understood the power of orthogonality in data analysis, we now review the orthogonal bio-kernel machine (Yang, 2005c).

Suppose a training peptide with R residues is denoted by $\mathbf{s}_n = (s_{n1}s_{n2}\cdots s_{nR}) \in \mathcal{A}^R$. A kernel peptide of the same length as the training peptide is denoted by $\boldsymbol{v}_h = (v_{h1}v_{h2}\cdots v_{hR}) \in \mathcal{A}^R$. The no-gap pairwise alignment score used in the original kernel space is shown in the following, where $m(a, b)$ is an alignment or mutation score from a to b:

$$\mathcal{K}(\boldsymbol{v}_h, \boldsymbol{s}_n) = \sum_{r=1}^{R} m(v_{hr}, s_{nr}) \tag{7.23}$$

Note that m can be any mutation matrix, such as the Dayhoff mutation matrix (Dayhoff and Schwartz, 1978) or the BLOSUM62 mutation matrix (Lipman *et al.*, 1989; Altschul *et al.*, 1990). An original kernel space with N training or testing peptides and H hypothetical kernel peptides is formulated as follows:

$$\mathbf{K} = \begin{pmatrix} k_{11} & k_{12} & \cdots & k_{1H} \\ k_{21} & k_{22} & \cdots & k_{2H} \\ \vdots & \vdots & \ddots & \cdots \\ k_{N1} & k_{N2} & \vdots & k_{NH} \end{pmatrix} \tag{7.24}$$

where the dataset is composed of N training peptides and H kernel peptides. Each entry of the matrix $\mathbf{K}$ is defined as follows:

$$k_{nh} = \mathcal{K}(\boldsymbol{v}_h, \boldsymbol{s}_n) \tag{7.25}$$

Exponential transformation was used to convert a kernel matrix $\mathbf{K}$ to a distance measure in the original orthogonal bio-kernel machine (Yang, 2005c). This transformation is shown in the following, where $\tilde{k}_{nh}$ measures the distance between $\boldsymbol{s}_n$ and $\boldsymbol{v}_k$, α is a constant and was set at ten:

$$\tilde{k}_{nh} = \exp\left(\alpha \frac{k_{nh} - \max(\mathbf{K})}{\max(\mathbf{K})}\right) \tag{7.26}$$

In the demonstration of the orthogonal bio-kernel machine in this book, this exponential transformation is replaced by mean–variance normalization, where $k_{\cdot,h}$ is a column vector of $\mathbf{K}$, $\overline{k}_{\cdot,h}$ is the mean of $k_{\cdot,h}$, and $\sigma(k_{\cdot,h})$ is the standard deviation of $k_{\cdot,h}$:

$$\tilde{k}_{\cdot,h} = \frac{k_{\cdot,h} - \overline{k}_{\cdot,h}}{\sigma(k_{\cdot,h})} \tag{7.27}$$

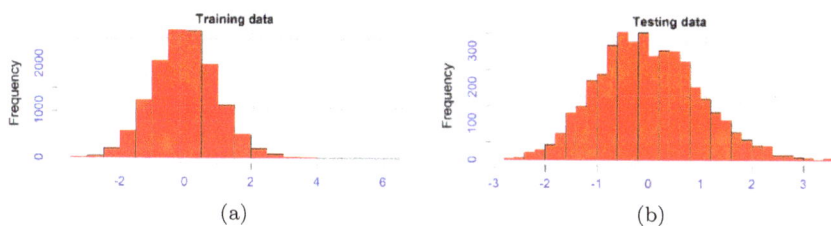

Figure 7.7. The densities of the training and testing datasets after the mean–variance normalization of the kernel matrix **K** for the O-linkage dataset. (a) The density of the training data. (b) The density of the testing data.

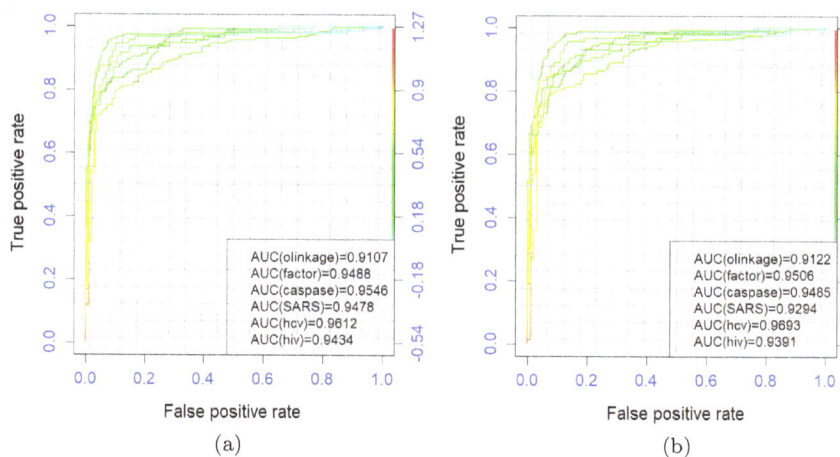

Figure 7.8. The ROC analysis of two bio-kernel spaces. (a) The LS bio-kernel machine model constructed on the mean–variance normalized data. (b) The LS bio-kernel machine model constructed on the mean–variance normalized data.

Figure 7.7 shows the outcome of this transformation for the O-linkage dataset. The density of the training dataset and the density of the testing dataset are similar.

Figure 7.8 shows the performance using ROC analysis for two sets of bio-kernel models. One was constructed in the original kernel space and the other was constructed in the mean–variance normalized kernel space. They do not show a significant difference.

After this transformation using the mean–variance normalization, **K** is transformed to **K̃**. The discussion of the orthogonal bio-kernel machine is based on $\tilde{\mathbf{K}}$. A linear regression model can be constructed in this bio-kernel space:

$$\mathbf{y} = \tilde{\mathbf{K}}\mathbf{w} + \mathbf{e} \tag{7.28}$$

where $\mathbf{y}$ is the label vector of the training peptides, $\mathbf{w} \in \mathcal{R}^H$ is a vector of model parameters, and $\mathbf{e}$ is an error vector. The $\tilde{\mathbf{K}}$ matrix is then expressed as kernel-oriented space, i.e., a collection of H column vectors corresponding to H hypothetical kernel peptides as shown in the following, where $\tilde{\mathbf{k}}_h$ corresponds to $\mathbf{v}_h$:

$$\tilde{\mathbf{K}} = (\tilde{\mathbf{k}}_1, \tilde{\mathbf{k}}_2, \ldots, \tilde{\mathbf{k}}_H) \tag{7.29}$$

The $\tilde{\mathbf{K}}$ matrix can be decomposed as the product of an orthogonal matrix $\mathbf{O}$ and a triangle matrix $\mathbf{T}$ as shown in the following:

$$\tilde{\mathbf{K}} = \mathbf{OT} \tag{7.30}$$

The orthogonal matrix $\mathbf{O}$ is also expressed as a collection of column vectors:

$$\mathbf{O} = (\mathbf{o}_1, \mathbf{o}_2, \ldots, \mathbf{o}_H) = \begin{pmatrix} o_{11} & o_{12} & \cdots & o_{1H} \\ o_{21} & o_{22} & \cdots & o_{2H} \\ \vdots & \vdots & \ddots & \cdots \\ o_{N1} & o_{N2} & \vdots & o_{NH} \end{pmatrix} \tag{7.31}$$

For the orthogonality operation on the bio-kernel space, the $\mathbf{T}$ matrix is designed as a triangular matrix with 1s as its diagonal entries. It is shown as follows:

$$\mathbf{T} = \begin{pmatrix} 1 & t_{12} & t_{13} & \cdots & t_{1,H-1} & t_{1H} \\ 0 & 1 & t_{23} & \cdots & t_{2,H-1} & t_{2H} \\ 0 & 0 & 1 & \cdots & t_{3,H-1} & t_{3H} \\ \vdots & \vdots & \vdots & \ddots & \vdots & \vdots \\ 0 & 0 & 0 & \cdots & 1 & t_{H-1,H} \\ 0 & 0 & 0 & \cdots & 0 & 1 \end{pmatrix} \tag{7.32}$$

The constraint applied to the orthogonal matrix $\mathbf{O}$ so that the product between the transpose of $\mathbf{O}$ and $\mathbf{O}$ results in a diagonal matrix $\mathbf{D}$ is shown in the following:

$$\mathbf{O}^t\mathbf{O} = \mathbf{D} \tag{7.33}$$

Each off-diagonal entry of the matrix $\mathbf{D}$ is shown as follows:

$$d_{ij} = \mathbf{o}_i^t\mathbf{o}_j = 0 \quad \forall i \neq j \tag{7.34}$$

Each diagonal entry of the matrix $\mathbf{D}$ is shown as follows:

$$d_{hh} = \mathbf{o}_h^t\mathbf{o}_h = \sum_{n=1}^{N} o_{nh}^2 \tag{7.35}$$

By using this orthogonality condition, the linear regression model defined earlier for the mean–variance normalized bio-kernel space is rewritten as follows:

$$\mathbf{y} = \mathbf{OTw} + \mathbf{e} = \mathbf{Og} + \mathbf{e} \tag{7.36}$$

It is normal to assume $\mathbf{e} \sim \mathcal{G}(0, 1)$. The solution to the previous model is thus defined as follows:

$$\mathbf{g} = (\mathbf{O}^t\mathbf{O})^{-1}\mathbf{O}^t\mathbf{y} = \mathbf{D}^{-1}\mathbf{O}^t\mathbf{y} \tag{7.37}$$

Because the matrix $\mathbf{D}$ is diagonal, it is easy to show that

$$\mathbf{D}^{-1} = \mathrm{diag}(d_{11}^{-1}, d_{22}^{-1}, \ldots, d_{HH}^{-1}) \tag{7.38}$$

Following the previous discussion, Eq. (7.38) is rewritten as

$$\mathbf{D}^{-1} = \mathrm{diag}((\mathbf{o}_1^t\mathbf{o}_1)^{-1}, \quad (\mathbf{o}_2^t\mathbf{o}_2)^{-1}, \ldots, (\mathbf{o}_H^t\mathbf{o}_H)^{-1}) \tag{7.39}$$

It has been shown in the abovementioned illustration that there is little or no correlation between variables (kernel peptides) in an orthogonal space. Therefore, the orthogonality operation generates a very simple expression of the $\mathbf{g}$ vector as shown in the following:

$$\mathbf{g} = \mathrm{diag}((\mathbf{o}_1^t\mathbf{o}_1)^{-1}, \quad (\mathbf{o}_2^t\mathbf{o}_2)^{-1}, \ldots, (\mathbf{o}_H^t\mathbf{o}_H)^{-1})\mathbf{O}^t\mathbf{y}$$
$$= \left(\frac{\mathbf{o}_h^t\mathbf{y}}{\mathbf{o}_h^t\mathbf{o}_h}\right)_{h=1}^{H} \tag{7.40}$$

It can be seen that each entry of **g** only relies on the corresponding column in the **O** matrix and has no relationship with any other columns, i.e.,

$$g_h \sim f(\mathbf{o}_h) \tag{7.41}$$

The relationship between **g**, **T**, and **W** is as shown previously, i.e.,

$$\hat{\mathbf{g}} = \mathbf{T}\mathbf{w} \tag{7.42}$$

Therefore, once **g** is determined, **w** can be estimated using the following equation:

$$\hat{\mathbf{w}} = (\mathbf{T}^t\mathbf{T})^{-1}\mathbf{T}^t\hat{\mathbf{g}} \tag{7.43}$$

Now, the question is how to estimate **g**. The common approach to solve this problem is an iterative process employing the Gram–Schmidt process (Bjorck, 1967; Chen *et al.*, 1991; Lyle and Trimble, 1991; Ward and Kincaid, 2009). This involves selecting columns in the orthogonal matrix **O** one by one. Two important criteria must be satisfied. First, every newly selected column $\mathbf{o}_h$ (hence a kernel peptide $\boldsymbol{v}_h$) must show the orthogonal relation with the previous ones. Second, the newly selected column must ensure maximized information gain. For a peptide discriminant problem, the information gain stands for the discrimination capability. The first kernel peptide selection approach employs the Fisher ratio. The first kernel peptide to be selected is thus based on the following criterion:

$$\mathbf{o}_1 = \mathbf{x}_\omega = \max_{1 \leq h \leq H} \left\{ \mathfrak{F}(\tilde{\mathbf{k}}_h, \mathbf{y}) \right\} \tag{7.44}$$

where $\mathfrak{F}(\tilde{\mathbf{k}}_h, \mathbf{y})$ is the Fisher ratio calculated for the hth column (hence the hth hypothetical kernel peptide) in the $\tilde{\mathbf{K}}$ matrix regarding the class label vector **y** defined as follows:

$$\mathfrak{F}(\mathbf{a}, \mathbf{y}) = \frac{|\mu_{\mathbf{a}, y=1} - \mu_{\mathbf{a}, y=2}|}{\sqrt{\sigma^2_{\mathbf{a}, y=1} + \sigma^2_{\mathbf{a}, y=2}}} \tag{7.45}$$

where **a** is an arbitrary numerical vector, $y = 1$ stands for a non-functional peptide, and $y = 2$ stands for a functional peptide. After the first selected kernel peptide, the matrix $\tilde{\mathbf{K}}$ is reorganized by swapping between two columns ($\tilde{\boldsymbol{k}}_1$ and $\tilde{\boldsymbol{k}}_\omega$). This swapping process moves $\tilde{\boldsymbol{k}}_\omega$ to the column of $\tilde{\boldsymbol{k}}_1$ and moves $\tilde{\boldsymbol{k}}_1$ to the column of $\tilde{\boldsymbol{k}}_\omega$. This swapping means that the hypothetical kernel $\tilde{\boldsymbol{k}}_\omega$ has been selected and the hypothetical kernel $\tilde{\boldsymbol{k}}_1$ is put back into the candidate list for later selection. At the same time, the hypothetical kernel $\tilde{\boldsymbol{k}}_\omega$ has been removed from the candidate list as it serves as the first best kernel in this orthogonal bio-kernel machine model. The new $\tilde{\mathbf{K}}$ matrix after this swapping is shown as follows:

$$\tilde{\mathbf{K}} = (\tilde{\boldsymbol{k}}_\omega, \tilde{\boldsymbol{k}}_2, \ldots, \tilde{\boldsymbol{k}}_{\omega-1}, \tilde{\boldsymbol{k}}_1, \tilde{\boldsymbol{k}}_{\omega+1}, \ldots, \tilde{\boldsymbol{k}}_K) \qquad (7.46)$$

The rest of the kernel peptides are selected using the following approach. First, the relation between $\mathbf{T}$, $\mathbf{O}$, and $\tilde{\mathbf{K}}$ matrices can be seen as follows:

$$\mathbf{T} = \mathbf{D}^{-1}\mathbf{O}^t\tilde{\mathbf{K}} \qquad (7.47)$$

The property of diagonal matrix $\mathbf{O}$ leads to the calculation of the entries of the $\mathbf{T}$ matrix:

$$t_{nh} = \frac{\mathbf{o}_n^t\tilde{\boldsymbol{k}}_h}{d_{hh}} \qquad (7.48)$$

The Gram–Schmidt process selects kernel peptides using the following equation based on the abovementioned relationship with the orthogonality constraint:

$$\mathbf{o}_h = \tilde{\boldsymbol{k}}_h - \sum_{i=1}^{h-1} t_{ih}\,\mathbf{o}_i \quad \forall h \in [2, H] \qquad (7.49)$$

When selecting the hth kernel peptide, there will be a number of candidates, i.e.,

$$\mathbf{o}_h^m = \tilde{\boldsymbol{k}}_m - \sum_{i=1}^{h-1} t_{ih}\,\mathbf{o}_i \quad \forall k \in [2, H] \; \delta \; m \in [h, H] \qquad (7.50)$$

For instance, the candidate set for selecting the second kernel peptide will have $H - 1$ candidates:

$$\mathcal{C}_2 = \{\mathbf{o}_2^2, \mathbf{o}_2^3, \ldots, \mathbf{o}_2^H\} \qquad (7.51)$$

The best candidate is selected from this list using the Fisher ratio defined earlier. After selection, a swapping process is carried out as was done for selecting the first kernel peptide shown earlier. The selection process will be terminated when the following inequality is satisfied, where ξ is a predefined threshold:

$$1 - \sum_{i=1}^{h} \overline{\overline{\mathfrak{F}}}(\mathbf{o}_i, \mathbf{y}) < \xi \tag{7.52}$$

where $\overline{\overline{\mathfrak{F}}}(\mathbf{o}_i, \mathbf{y})$ is defined as follows:

$$\overline{\overline{\mathfrak{F}}}(\mathbf{o}_i, \mathbf{y}) = \frac{\mathfrak{F}(\mathbf{o}_i, \mathbf{y})}{\sum_{g=1}^{h_\pi} \mathfrak{F}(\tilde{\mathbf{k}}_g, \mathbf{y})} \tag{7.53}$$

Figure 7.9 shows the ROC curves as well as the AUC values for the orthogonal bio-kernel models constructed for six peptide datasets using both the Dayhoff mutation matrix and the BLOSUM62 mutation matrix. The termination condition (ξ) was set to 0.001. All models performed well and the difference between these two sets of models was insignificant. The left part of Table 7.1 shows all the

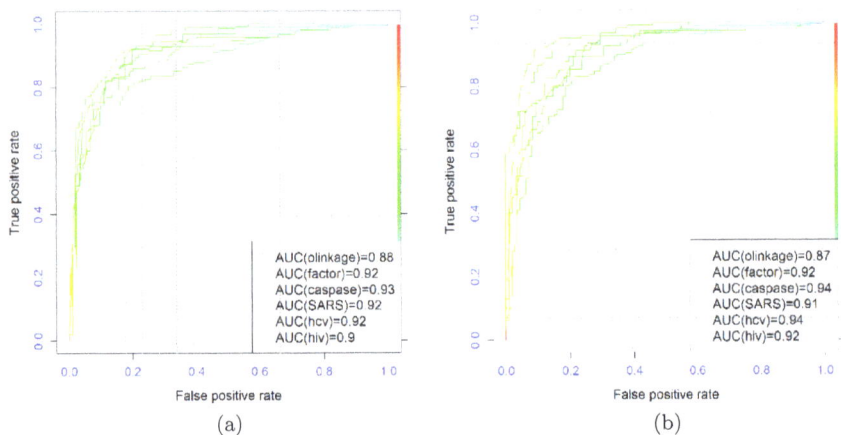

Figure 7.9. The ROC analysis of two orthogonal bio-kernel models constructed for six peptide datasets using two mutation matrices with the Fisher ratio for the selection of the orthogonal kernels. (a) The models constructed using the Dayhoff mutation matrix. (b) The models constructed using the BLOSUM62 mutation matrix.

Table 7.1. The AUC and MCC values for the orthogonal bio-kernel models constructed for six peptide datasets using the Fisher test approach and the AUC test approach for the selection of the orthogonal kernels. Dayhoff stands for the models constructed using the Dayhoff mutation matrix and BLOSUM62 stands for the models constructed using the BLOSUM62 mutation matrix.

| | Fisher test approach | | | | AUC test approach | | | |
| | Dayhoff | | BLOSUM62 | | Dayhoff | | BLOSUM62 | |
Data name	AUC	MCC	AUC	MCC	AUC	MCC	AUC	MCC
O-linkage	0.88	0.59	0.87	0.53	0.85	0.45	0.87	0.58
Factor Xa	0.92	0.50	0.92	0.55	0.79	0.25	0.91	0.57
Caspase	0.93	0.67	0.94	0.76	0.88	0.58	0.96	0.84
SARS	0.92	0.68	0.91	0.62	0.91	0.68	0.94	0.73
HCV	0.92	0.41	0.94	0.41	0.92	0.42	0.95	0.40
HIV	0.90	0.58	0.92	0.61	0.90	0.57	0.91	0.61

Table 7.2. The number of bio-kernels employed in the final orthogonal bio-kernel models constructed for six peptide datasets using the Fisher test approach and the AUC test approach for the selection of the orthogonal kernels. Dayhoff stands for the models constructed using the Dayhoff mutation matrix and BLOSUM62 stands for the models constructed using the BLOSUM62 mutation matrix.

| | Fisher test approach (%) | | AUC test approach (%) | |
Data name	Dayhoff	BLOSUM62	Dayhoff	BLOSUM62
O-linkage	2.64	29.18	54.08	20.91
Factor Xa	1.89	1.54	37.73	24.7
Caspase	3.81	4.23	64.24	61.8
SARS	4.11	4.06	74.85	50.92
HCV	0.87	0.84	17.2	7.43
HIV	2.74	17.99	31.68	49.67

numerical values of the performance for the two sets of the models, which show a slight difference in the individual datasets. Table 7.2 shows the number of kernels employed by the orthogonal bio-kernel models. It can be seen that less than 30% of bio-kernels were employed in these models, but good performance is still maintained for all. Most models employed less than 5% of bio-kernels. In the orthogonal bio-kernel models constructed for the HIV dataset, less

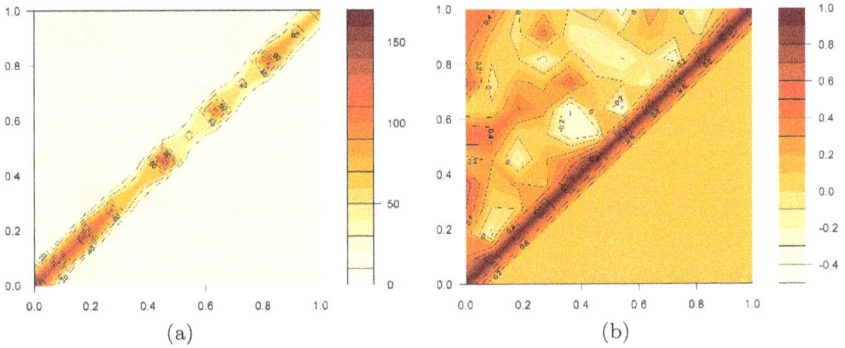

Figure 7.10. The properties of the **O** and **T** matrices for the orthogonal bio-kernel models constructed for the O-linkage dataset. (a) The contour of the $\mathbf{O}^t\mathbf{O}$ matrix. (b) The contour of the **T** matrix.

than 1% of bio-kernels were employed to achieve a satisfactory and comparable performance.

Figure 7.10 shows the properties of the **O** and **T** matrices derived through orthogonal learning of the bio-kernel models for the O-linkage dataset. The contour on the left panel shows the property of $\mathbf{O}^t\mathbf{O}$. It can be seen that the matrix $\mathbf{O}^t\mathbf{O}$ is indeed diagonal; hence, **O** is indeed orthogonal. The contour on the right demonstrates the derived **T** matrix. It shows that **T** is indeed a triangular matrix.

This discussion introduced the use of the Fisher test approach for constructing an orthogonal bio-kernel model, i.e., the termination condition is based on the Fisher ratio test. The next approach employs the AUC test approach for constructing an orthogonal bio-kernel model. By using this approach, optimal bio-kernels are selected in terms of their contributions to the model performance, i.e., their AUC values. Suppose the AUC of a bio-model employing h orthogonal bio-kernels is denoted by $\Gamma(\mathbf{O}_h)$, where $\mathbf{O}_h$ represents a matrix till column h. Suppose $h-1$ kernels have been well determined through this orthogonalization process and we need to search for the hth orthogonal kernel. We then build up a candidate list, which is a collection of all the $H - h + 1$ kernels which have not yet been employed. A candidate list is thus such a set:

$$\mathcal{C}_h = \{\Gamma^h(\mathbf{O}_{h-1}, \mathbf{o}_h^h), \Gamma^{h+1}(\mathbf{O}_{h-1}, \mathbf{o}_h^{h+1}), \ldots, \Gamma^H(\mathbf{O}_{h-1}, \mathbf{o}_h^H)\} \quad (7.54)$$

One kernel is selected from this candidate set C_h to satisfy the following condition, where $\mathbf{o}_h^+$ is a new orthogonal kernel to be included into the orthogonal matrix $\mathbf{O}$:

$$\mathbf{o}_h^+ = \underset{h}{\operatorname{argmax}}\{C_h\} \tag{7.55}$$

The termination condition is also revised and is shown as follows:

$$1 - \bar{\bar{\Gamma}}(\mathbf{O}_h) < \xi \tag{7.56}$$

where $\bar{\bar{\Gamma}}(\mathbf{O}_h)$ is defined as follows:

$$\bar{\bar{\Gamma}}(\mathbf{O}_h) = \frac{\Gamma(\mathbf{O}_h)}{\Gamma(\mathbf{O}_{h-1})} \tag{7.57}$$

The right panel of Table 7.1 shows the performance, where the orthogonal bio-kernel models constructed using the AUC test approach demonstrate comparable performance to the models constructed using the Fisher ratio test approach, especially when using the BLOSUM62 mutation matrix for establishing a kernel matrix. The right part of Table 7.2 shows the proportion of the bio-kernels employed in these models. The difference is that these models employ significantly more bio-kernels.

7.3 Bayesian bio-kernel machine

The Bayesian bio-kernel machine is another extension of the bio-kernel machine (Yang, 2005b) with the aim of enhancing the generalization capability of the bio-kernel machine model. For convenience, several notations used earlier have been slightly revised in this section. A bio-kernel machine with H hypothetical bio-kernels is denoted by the following equation:

$$\hat{y}_n = \sum_{h=1}^{H} w_h \mathcal{K}(\boldsymbol{v}_h, \mathbf{s}_n) = y_n - e_n \tag{7.58}$$

where $\mathbf{s}_n$ is the nth training peptide, $\boldsymbol{v}_h$ is a peptide which serves as the hth hypothetical bio-kernel, w_h is the weight for the hth hypothetical bio-kernel, $\mathcal{K}(a,b)$ is the kernel function which employs a mutation matrix to measure the mutation probability from peptide

a to peptide b (hence the similarity between a and b), y_n and $\hat{y}_n$ are the observed class label and the predicted class label of the known peptide $\mathbf{s}_n$, respectively, and e_n is the error between y_n and $\hat{y}_n$.

In the Bayesian bio-kernel machine, the error is defined as the abovementioned Gaussian distribution. According to the convention, α^{-1} is used to replace σ_e^2 in the error function shown as follows:

$$e_n \sim \mathcal{G}(0, \sigma_e^2) = \mathcal{G}(0, \alpha^{-1}) \tag{7.59}$$

In addition, a super distribution of α is defined as follows:

$$\alpha \sim \mathcal{G}(0, \vartheta) \tag{7.60}$$

The Gaussian distribution is also applied to the model parameters $\mathbf{w}$ as in most Bayesian studies (Bullen *et al.*, 2003):

$$w_h \sim \mathcal{G}(0, \beta^{-1}) \tag{7.61}$$

where $\beta^{-1} = \sigma_w^2$. Another super distribution is applied to β:

$$\beta \sim \mathcal{G}(0, \tau) \tag{7.62}$$

Suppose a single hyper-parameter ω is used as a collection of both ϑ and τ. In Bayesian learning, the relationship between the notations of the abovementioned parameters is defined as follows:

$$p(\mathbf{w}, \omega | \mathbf{y}) = \frac{p(\mathbf{y} | \mathbf{w}, \omega) p(\mathbf{w}, \omega)}{p(\mathbf{y})} \tag{7.63}$$

where $p(\mathbf{y} | \mathbf{w}, \omega)$ is the conditional probability, which is also referred to as the likelihood, $p(\mathbf{y})$ is referred to as the normalization factor, which is also called the evidence, and $p(\mathbf{w}, \omega)$ is referred to as the prior structure. This prior structure is decomposed to two terms as shown in the following:

$$p(\mathbf{w}, \vartheta) = p(\mathbf{w} | \vartheta) p(\omega) \tag{7.64}$$

In Eq. (7.64), $p(\mathbf{w} | \omega)$ is referred to as the conditional probability and $p(\omega)$ is defined as the *a priori* probability of all the hyper-parameters. Thus, the posterior probability $\mathscr{P}$ is simplified as the multiplication between $p(\mathbf{y}|\mathbf{w}, \omega)$, $p(\mathbf{w}|\omega)$, and $p(\omega)$. It is shown as follows:

$$\mathscr{P} \propto p(\mathbf{y}|\mathbf{w}, \omega)p(\mathbf{w}|\omega)p(\omega) \qquad (7.65)$$

The evidence $p(\mathbf{y})$ can be ignored because it is a constant in a constructed model. Now, the three abovementioned terms are shown in the following, where $\tilde{\mathbf{K}} \in \mathcal{R}^{N \times H}$ is a kernel matrix:

$$p(\mathbf{y}|\mathbf{w}, \omega) = \left(\frac{\alpha}{2\pi}\right)^{\frac{N}{2}} \exp\left(-\frac{\alpha}{2}(\tilde{\mathbf{K}}\mathbf{w} - \mathbf{y})^t(\tilde{\mathbf{K}}\mathbf{w} - \mathbf{y})\right) \qquad (7.66)$$

and

$$p(\mathbf{w}|\omega) = \left(\frac{\beta}{2\pi}\right)^{\frac{H}{2}} \exp\left(-\frac{\beta}{2}\mathbf{w}^t\mathbf{w}\right) \qquad (7.67)$$

as well as

$$p(\omega) = \left(\frac{\vartheta}{2\pi}\right)^{1/2} \exp\left(-\frac{\vartheta}{2}\alpha^2\right) \left(\frac{\tau}{2\pi}\right)^{1/2} \exp\left(-\frac{\tau}{2}\beta^2\right) \qquad (7.68)$$

An even more simplified form of the model can be brought about by applying the negative logarithm to $\mathscr{P}$. It is shown as follows:

$$\begin{aligned}
&-\log(\mathscr{P}) \\
&= \frac{1}{2}\left\{ \begin{array}{c} \alpha(\tilde{\mathbf{K}}\mathbf{w} - \mathbf{y})^t(\tilde{\mathbf{K}}\mathbf{w} - \mathbf{y}) + \beta\mathbf{w}^t\mathbf{w} + \vartheta\alpha^2 + \tau\beta^2 \\ -N\log\alpha - H\log\beta - \log\vartheta - \log\tau \end{array} \right\}
\end{aligned} \qquad (7.69)$$

One of the most used approaches to solving this model is the maximum a posterior (MAP) approach (Mitchell, 1997), which has been employed for constructing a Bayesian bio-kernel machine model (Yang, 2005b). With this approach, the model parameters ($\mathbf{w}$, ϑ, and τ) are updated in an iterative process. The hyper-parameter α is estimated using the following equation based on MAP:

$$\alpha = \frac{-\mathbf{e}^2 + \sqrt{\mathbf{e}^4 + 8\vartheta N}}{4\vartheta} \qquad (7.70)$$

where $\mathbf{e} \in (e_1, e_2, \dots, e_N)$ is an error vector shown as follows:

$$\mathbf{e} = \mathbf{y} - \hat{\mathbf{y}} \qquad (7.71)$$

The hyper-parameter β is estimated using the following equation based on the same procedure:

$$\beta = \frac{-\mathbf{w}^t\mathbf{w} + \sqrt{(\mathbf{w}^t\mathbf{w})^2 + 8\tau H}}{4\tau} \tag{7.72}$$

Based on the estimated hyper-parameters (α and β), model parameter vector $\mathbf{w}$ is estimated using the following equation:

$$\hat{\mathbf{w}} = \alpha(\alpha\tilde{\mathbf{K}}^t\tilde{\mathbf{K}} + \beta\mathbf{I})^{-1}\tilde{\mathbf{K}}^t\mathbf{y} \tag{7.73}$$

Finally, the super hyper-parameters are updated using the following equations:

$$\vartheta = \alpha^{-2} \tag{7.74}$$

and

$$\tau = \beta^{-2} \tag{7.75}$$

The Bayesian bio-kernel machine is thus applied to six peptide datasets using both the Dayhoff and BLOSUM62 mutation matrices to measure the similarity between peptides. Table 7.3 shows the AUC and MCC measurements for these two sets of models. Figure 7.11 shows their ROC curves. Two sets of models demonstrate similar performance.

Figure 7.12 shows the distributions of four key hyper-parameters during the construction of the Bayesian bio-kernel machine model

Table 7.3. The performance (AUC and MCC) of the Bayesian bio-kernel models constructed for six peptide datasets. Dayhoff stands for the models constructed using the Dayhoff mutation matrix and BLOSUM62 stands for the models constructed using the BLOSUM62 mutation matrix.

Data name	Dayhoff		BLOSUM62	
	AUC	MCC	AUC	MCC
O-linkage	0.91	0.69	0.90	0.62
Factor Xa	0.87	0.48	0.92	0.57
Caspase	0.93	0.73	0.96	0.82
SARS	0.90	0.60	0.90	0.66
HCV	0.95	0.42	0.94	0.42
HIV	0.89	0.56	0.80	0.51

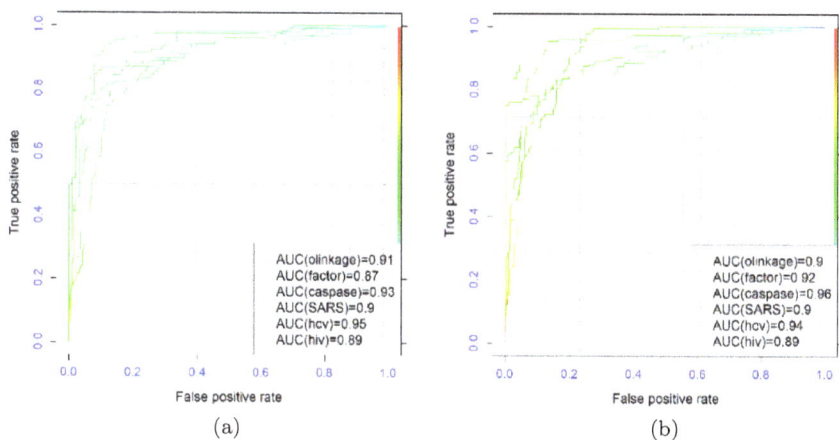

Figure 7.11. The performance evaluation of the Bayesian bio-kernel machine models constructed for six peptide datasets. (a) The models constructed using the Dayhoff mutation matrix. (b) The models constructed using the BLOSUM62 mutation matrix.

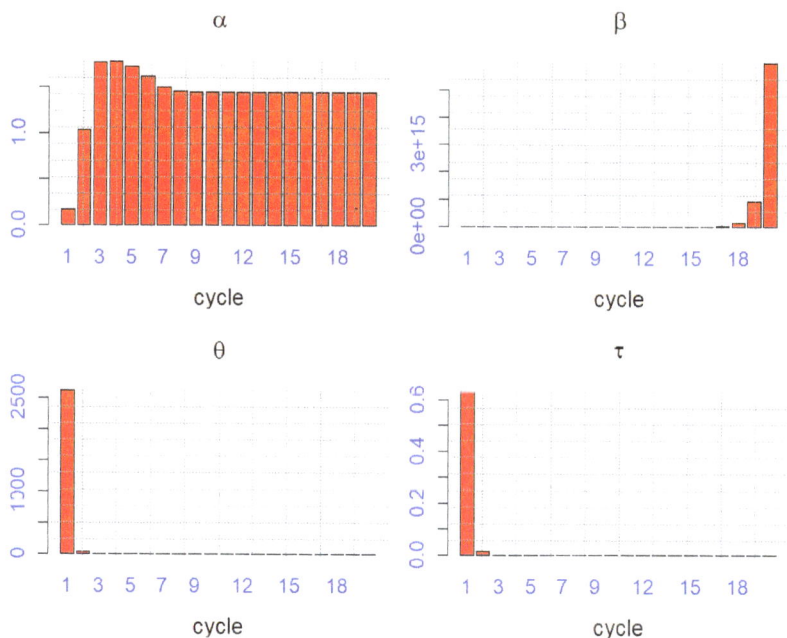

Figure 7.12. The distribution of four key hyper-parameters (α, β, θ, and τ) of the Bayesian bio-kernel model constructed for the O-linkage dataset. Cycle stands for the learning iterations.

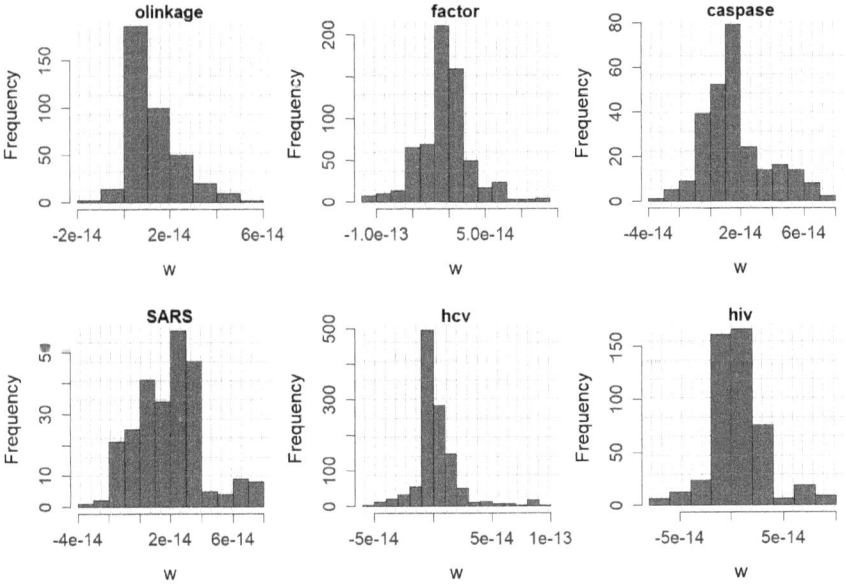

Figure 7.13. The densities of the weights of the Bayesian bio-kernel models constructed for six peptide datasets.

for the O-linkage dataset. The distributions show that these hyper-parameters converge very quickly during a Bayesian learning process.

Figure 7.13 shows the weight distributions of the Bayesian bio-kernel machine models constructed for six peptide datasets. They generally show normal distributions.

7.4 Intelligent bio-kernel machines

The abovementioned bio-kernel machines are based on machine learning algorithms which generate quantitative models for predictions. For instance, a model shown in the following will generate a numerical and quantitative prediction $\hat{y}_n \in \mathcal{R}$ for each numerical input:

$$\mathbf{y} = \tilde{\mathbf{K}}\mathbf{w} \tag{7.76}$$

where $\tilde{\mathbf{K}} = (\tilde{\boldsymbol{k}}_{nh})_{n=1,h=1}^{N,H} \in \mathcal{R}^{N \times H}$ and

$$\tilde{\boldsymbol{k}}_n = \left(\tilde{\mathcal{K}}(\boldsymbol{v}_1, \mathbf{s}_n), \tilde{\mathcal{K}}(\boldsymbol{v}_2, \mathbf{s}_n), \ldots, \tilde{\mathcal{K}}(\boldsymbol{v}_H, \mathbf{s}_n)\right) \tag{7.77}$$

A decision-making process follows the generation of a numerical prediction $\hat{y}_n$ for $\tilde{k}_n$, i.e., s_n is either a functional peptide or a non-functional peptide. Such a quantitative model may not deliver a qualitative inference such as the reasoning rule shown in the following:

$$s_n \text{ is } \begin{cases} \text{cleaved} & \text{If an Arginine appears at } P_1 \\ \text{non--cleaved} & \text{otherwise} \end{cases} \tag{7.78}$$

Or, we can have another reasoning rule shown as follows:

$$s_n \text{ is } \begin{cases} \text{cleaved} & \text{because it is very similar to IEGRI} \\ \text{non--cleaved} & \text{otherwise} \end{cases} \tag{7.79}$$

where $s_n = $ IEGRT is a peptide waiting for the identification of its cleavage status, i.e., whether it is a cleaved Factor Xa peptide or not. Meanwhile, IEGRI is a known cleaved Factor Xa peptide and is served as a kernel peptide in models.

Exploring qualitative intelligence from data is one of the most thriving areas in machine learning and has been exercised in the machine learning community for a very long time. The most important feature of this kind of model is that the intelligence explored from data can be expressed in a similar way as the human reasoning process. The machine learning algorithms employed for qualitative intelligence exploration from data are mainly decision tree algorithms.

One of the important properties of decision tree algorithms is that they can work in a high-dimensional space and recognize the contributing variables, such as kernel peptides, in a bio-kernel machine model. Two representative algorithms are the decision tree algorithm (ID3) (Quinlan, 1986) and the classification and regression tree algorithm (CART) (Breiman *et al.*, 1984). Both algorithms employ the "divide-and-conquer" principle (Karatsuba and Ofman, 1963) in a learning process. The "divide-and-conquer" principle is in fact a binary search process (Shapiro and Haralick, 1982). With this search strategy, a decision-making space is divided into two halves. The truth for a question is in either one of two halves, i.e., either one of two subspaces. A decision-making space is gradually divided until

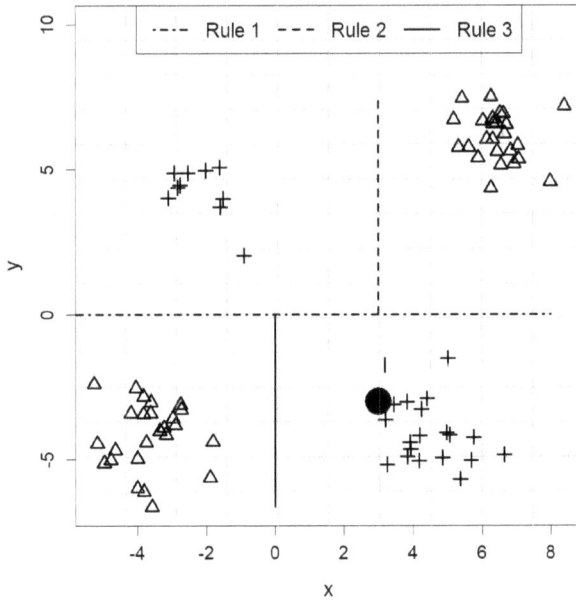

Figure 7.14. An illustration of the rule of divide and conquer. Two classes of points are labeled by the triangles and the pluses. The filled dot is an unknown point for predicting its class label through a learning process.

a subspace can definitely answer a question, i.e., the subspace where a question is located only contains one class of data points.

Figure 7.14 gives a simple demonstration in a two-dimensional space, where two classes of data points are labeled by the triangles and the pluses. Without a division, this space is mixed by two classes of data points and it is impossible to classify the question data point, which is marked by a filled dot correctly using a single hyperplane (a straight line) which is perpendicular to a coordinate (a variable). To label the question data point according to the class in which it belongs, a learning process must be employed, by which the space is divided into four subspaces by introducing three divisions and hence three partitioning rules. The first partitioning rule ($y = 0$) certainly cannot label the question data point correctly. The second partitioning rule ($x = 3$) still cannot label the question data point correctly. The third partitioning rule ($x = 0$) divides the lower subspace into two even smaller sub-subspaces. In the right

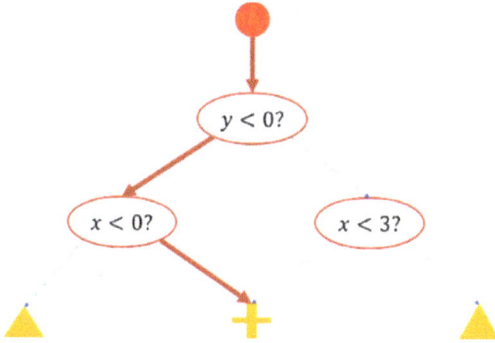

Figure 7.15. The decision-making tree for the hierarchical organization of three partitioning rules according to the data shown in Figure 7.14.

sub-subspace, the question data point can be labeled with confidence because all data points in this sub-subspace are pluses.

Figure 7.15 shows a decision tree of a hierarchical organization of three induced partitioning rules for the data shown in Figure 7.14. Within this hierarchical inference model, the decision is made by traveling from the root of the tree to one of the bottoms (a leaf) through a number of comparisons, such as the comparison at the root node $y < 0$ and the comparison at one branch node $x < 0$. Suppose we need to label the question data point. The first comparison is to examine whether the filled dot is located below or above the partitioning rule $y = 0$ in this two-dimensional space. Because the lower space of the partitioning rule $y = 0$ is composed of two classes of known data, a further comparison is made against the partitioning rule $x = 0$. The question data point is located in the right subspace generated by $x = 0$ of the lower subspace generated by $y = 0$. The question data point marked by the filled dot is thus labeled as the plus class.

In order to automate a learning process for a decision tree model, a quantitative metric must be employed. The two most commonly employed quantitative metrics are the Gini index (Breiman *et al.*, 1984) and the information gain (Breiman *et al.*, 1984; Quinlan, 1986). The Gini index is defined as follows:

$$I_G(x = \tau) = \sum_{k=1}^{K} p_k(\tau)[1 - p_k(\tau)] \tag{7.80}$$

where K is the number of classes in a dataset $\mathcal{D}$, $x \in V$ is a variable of the variable set V, $x = \tau$ is a partitioning rule, and $p_k(\tau)$ is the probability or proportion of data points in the subspace generated by the partitioning rule $x = \tau$ of the kth class. For instance, in Figure 7.14, there are 45 data points in the lower subspace generated by the partitioning rule $y = 0$, in which 25 are triangles and 20 are pluses. The probability that the question data point (the filled dot) belongs to the triangle class is thus $25/45$ and the probability that the question data point belongs to the plus class is $20/45$. However, after employing the third partitioning rule $x = 0$, the probability that the question data point belongs to the plus class is 100%. In Eq. (7.80), the product $p_k(\tau)[1 - p_k(\tau)]$ is referred to as the Bernoulli variance. For instance, the Bernoulli variance of the lower subspace in Figure 7.14 is

$$I_G(y = 0) = \frac{25}{45}\left(1 - \frac{25}{45}\right) + \frac{20}{45}\left(1 - \frac{20}{45}\right) = 0.4938 \qquad (7.81)$$

The Bernoulli variance of the bottom-right subspace in Figure 7.14 is

$$I_G(y = 0) = \frac{20}{20}\left(1 - \frac{0}{20}\right) + \frac{0}{20}\left(1 - \frac{20}{20}\right) = 1 \qquad (7.82)$$

Therefore, the Bernoulli variance is one if a subspace is pure for one class of data points and is approaching 0.5 $(0.25 \times K)$ if a subspace is composed of similar number of data points from two classes. The information gain is calculated based on the entropy theory, which is defined as follows:

$$I_E(x = \tau) = \sum_{k=1}^{K} p_k(\tau)\log_2(1 + p_k(\tau)) \qquad (7.83)$$

In Eq. (7.83), $p_k(\tau)$ has the same meaning as the Gini index, and replacing $\log_2 p_k(\tau)$ by $\log_2(1 + p_k(\tau))$ is done to avoid an infinite value once $p_k(\tau)$ approaches zero. In a learning process, the information gain must be maximized when evaluating a partitioning rule because the information gain will be one if a subspace is pure for one class of data points. In this case, $p_k(\tau) = 1$; hence,

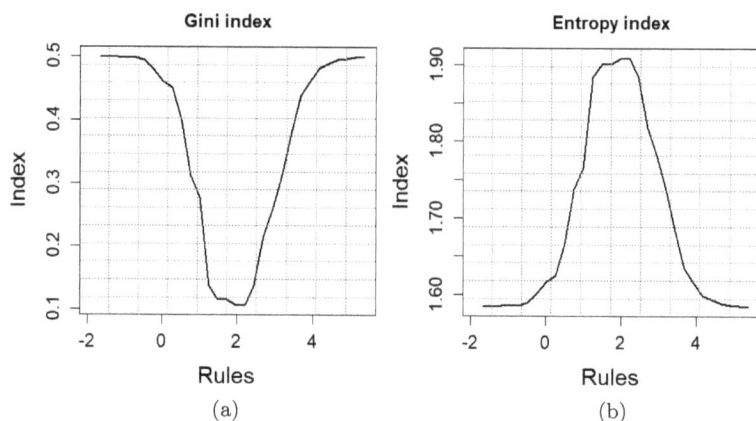

Figure 7.16. The impurity measurement for the partitioning rules applied to the data shown in Figure 3.36(a). (a) The Gini index (b) The entropy index (information gain). Adapted from Yang (2022a).

$p_k(\tau)\log_2(1 + p_k(\tau)) = 1$ and $\max\{I_E(x = \tau)\} = K$. Figure 7.16(b) shows the distribution of the Gini index and the information gain.

CART partitions a space hierarchically to explore inference or partitioning rules based on the Gini index (Breiman *et al.*, 1984). The algorithm grows a tree through a process which hierarchically and consecutively divides a data space. The partition process employs a set of abovementioned partitioning rules. The process is terminated once the Gini index turns to zero and the bottom nodes (leaves) are labeled by the class of the data points within a subspace with 100% confidence. Afterward, a grown tree is pruned, aiming to generate a better generalization performance.

Figure 7.17 shows an example of a division process employing six partitioning rules hierarchically in a two-dimensional space. The last three partitioning rules (Rule 4, Rule 5, and Rule 6) generate subspaces with a few data points. Figure 7.18(a) shows an overgrown tree by employing these six partitioning rules. In fact, the four data points in the last three subspaces are perhaps the outliers and the last three partitioning rules may therefore reduce the generalization performance.

Figure 7.18(a) shows a possibly overgrown tree for the data shown in Figure 7.17, in which the Gini index values in all leaf

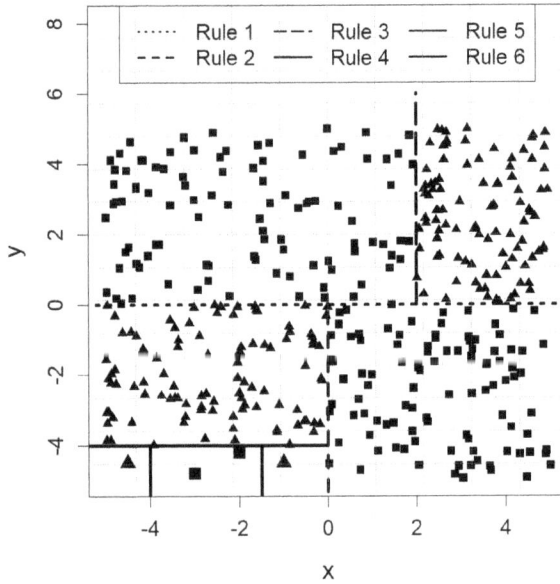

Figure 7.17. An example of too many divisions, which may lead to an overgrown tree.

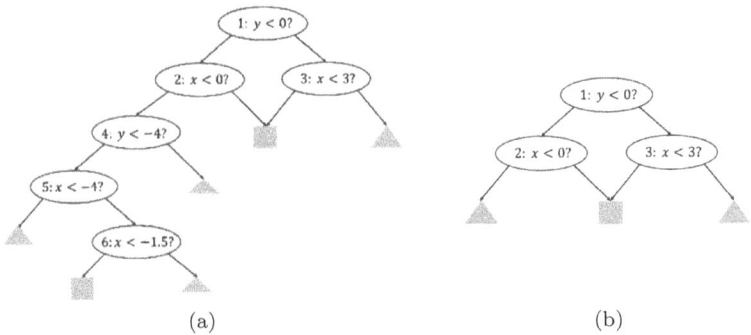

(a)   (b)

Figure 7.18. (a) An overgrown tree for the divisions made in the data space shown in Figure 7.17. (b) A pruned tree.

nodes are zero. However, such a tree or model may not have good generalization capability because the three subspaces on the bottom left of Figure 7.17 may be the space of the triangle class and the two squares may be the outliers. The second step of CART prunes this kind of node by testing the generalization performance. Any

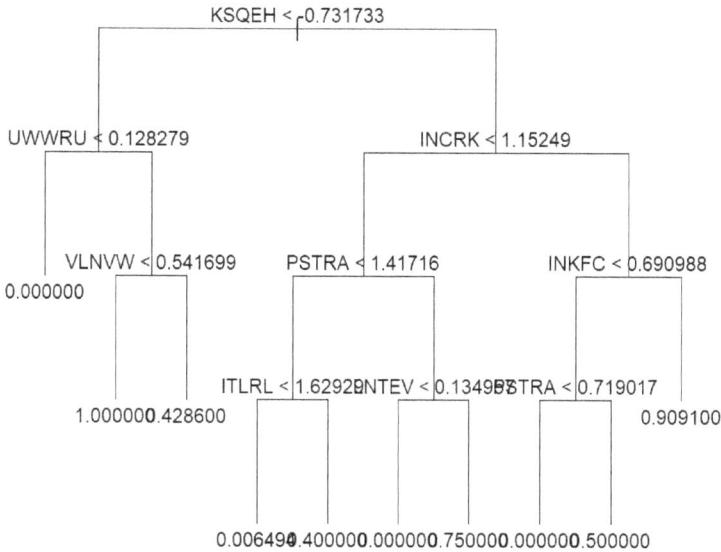

Figure 7.19. A CART bio-kernel model constructed for the Factor Xa dataset.

node which may decrease the generalization performance is pruned. After pruning, a tree with three nodes, as shown in Figure 7.18(b), may have better generalization performance. Figure 7.19 shows a tree generated by CART for the Factor Xa dataset. In this tree, it can be seen that different kernel peptides play different roles in this decision-making process, i.e., they are located at different nodes including the root node in this hierarchical structure. The kernel peptide KSQEH is a peptide which has the maximum Gini index so that it is located at the root node.

Figure 7.20 shows an investigation of the relationship between the number of the selected kernel peptides and the model discrimination capability in a CART model constructed for the Factor Xa dataset. Figure 7.20(a) shows the distribution of the t test p values of each kernel peptide selected during the process of growing a tree as shown in Figure 7.19. In total, nine kernel peptides were selected to grow a CART model. The t test p values of the nine models show a trend of reducing the significance of the selected kernel peptides in the grown tree as shown in Figure 7.19. For instance, the kernel peptide selected at the root node (KSQEH) has the smallest p value but

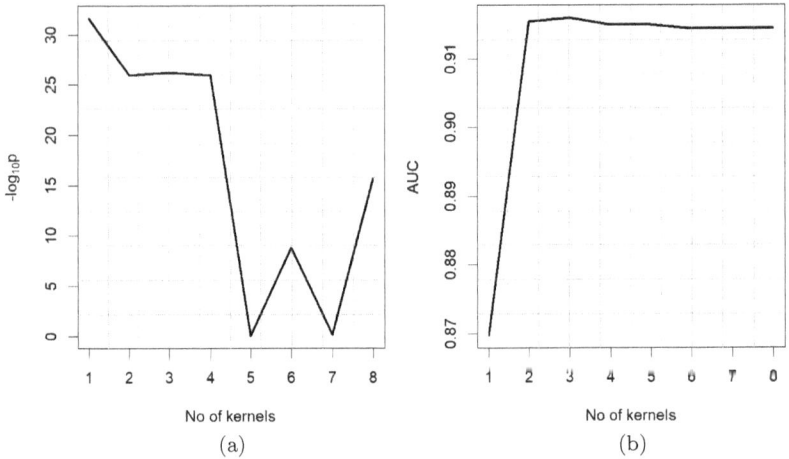

Figure 7.20.   The relationship between the number of the kernel peptides and the model discriminant capability in the CART bio-kernel models constructed for the Factor Xa dataset. (a) The distribution of the t-test p values. (b) The distribution of AUC values.

the kernel peptides selected later such as UWWRU, VLNVW, and ITLRL have greater and greater p values, i.e., the p value of the t test for VLNVW is greater than that for UWWRU and the p value of the t test for ITLRL is greater than that for VLNVW, etc. In contrast, the nine LS bio-kernel models constructed on the selected kernel peptides show a trend of pushing the discrimination capability (AUC) up. Note that the first LS bio-kernel machine model was constructed based on the first selected kernel peptide, i.e., KSQEH as shown in Figure 7.19. The second LS bio-kernel machine model was constructed based on the first two selected kernel peptides, i.e., KSQEH and UWWRU.

Table 7.4 shows the performance evaluation of the CART bio-kernel machine models constructed for six peptide datasets. Figure 7.21 shows their ROC curves. It can be seen that the performance is not as good as previous bio-kernel machine models. The AUC values of several models were even below 0.8. This is as expected because an intelligent bio-kernel model does not employ all the hypothetical kernel peptides. It only employs a small proportion

Table 7.4. The performance (AUC and MCC) evaluation of the CART bio-kernel machine models constructed for six peptide datasets. Dayhoff stands for the models constructed using the Dayhoff mutation matrix and BLOSUM62 stands for the models constructed using the BLOSUM62 mutation matrix.

Data name	Dayhoff		BLOSUM62	
	AUC	MCC	AUC	MCC
O-linkage	0.80	0.42	0.77	0.43
Factor Xa	0.75	0.31	0.83	0.44
Caspase	0.89	0.68	0.87	0.71
SARS	0.84	0.59	0.86	0.59
HCV	0.91	0.51	0.94	0.51
HIV	0.83	0.53	0.76	0.47

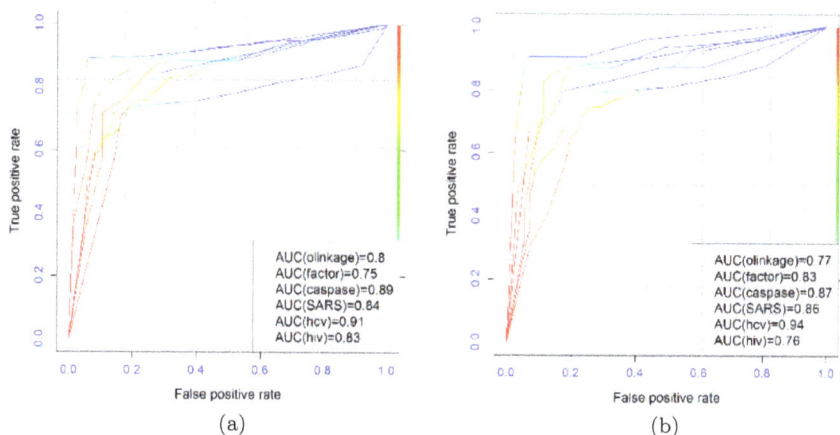

Figure 7.21. The ROC curves of the CART bio-kernel machine models constructed for six peptide datasets. (a) The ROC curves of the models using the Dayhoff matrix. (b) The ROC curves of the models using the BLOSUM62 matrix.

of the available hypothetical kernel peptides. The important issue is that the selection does not consider the correlations between the hypothetical kernel peptides. Both the Gini index and the information gain are only used when designing a partitioning rule for each individual kernel peptide. Moreover, the performance of these

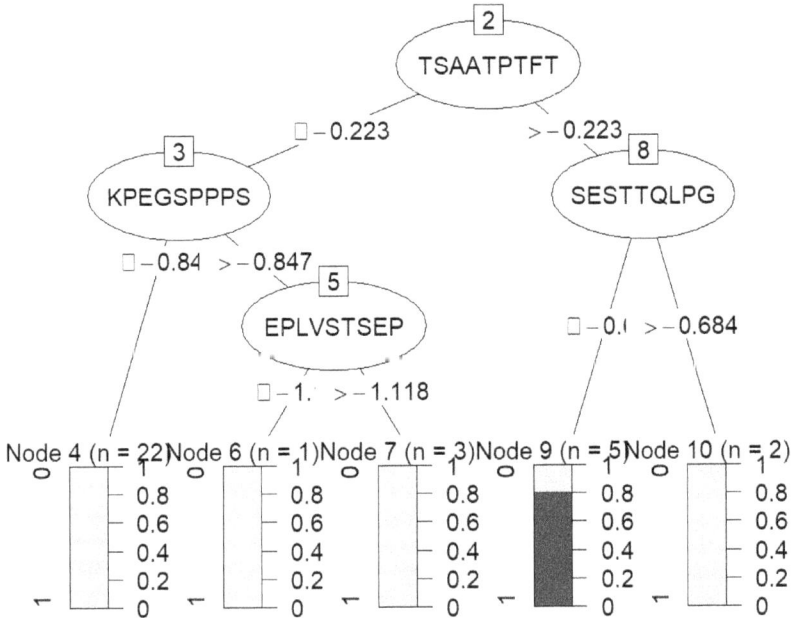

Figure 7.22. The C5.0 bio-kernel model for the O-linkage dataset.

models is also sacrificed for the parsimonious model structures. The CART algorithm and the C5.0 algorithm, which will be discussed later, partition a space using orthogonal partitioning rules, i.e., each partitioning rule is perpendicular to a coordinate. For instance, partitioning rules $y = 0$ and $x = 0$, as shown in Figure 7.14. This issue will be discussed when introducing the random forest algorithm later in this chapter.

Another promising algorithm is the decision tree algorithm called the Iterative Dichotomiser 3 (ID3) (Quinlan, 1986). ID3 was upgraded to C4.5 and C5.0 afterward (Quinlan, 1993). The metric used in C5.0 for designing partitioning rules when growing a decision tree is the information gain. Figure 7.22 shows the intelligent bio-kernel model (a tree) constructed by C5.0 for the O-linkage dataset. Again, the model structure is parsimonious where only four kernel peptides were employed.

Table 7.5. The performance evaluation of the C5.0 bio-kernel models constructed for six peptide datasets. Dayhoff stands for the models constructed using the Dayhoff mutation matrix and BLOSUM62 stands for the models constructed using the BLOSUM62 mutation matrix.

Data name	Dayhoff		BLOSUM62	
	AUC	MCC	AUC	MCC
O-linkage	0.75	0.50	0.78	0.56
Factor Xa	0.78	0.55	0.77	0.56
Caspase	0.84	0.69	0.89	0.78
SARS	0.82	0.64	0.86	0.72
HCV	0.9	0.78	0.88	0.75
HIV	0.83	0.65	0.78	0.55

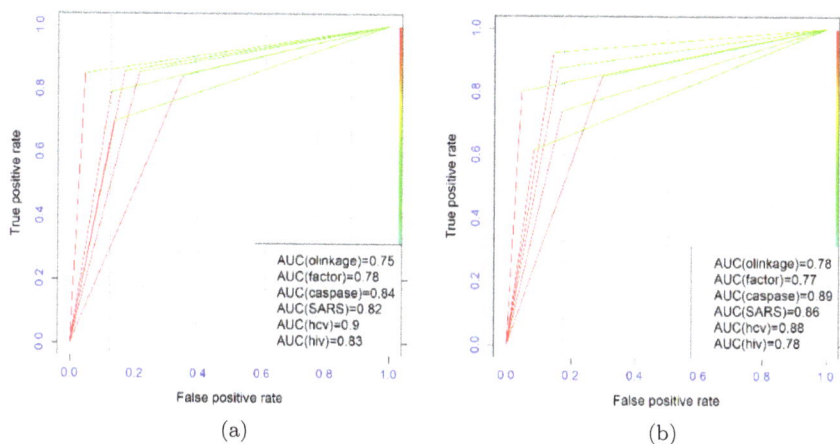

Figure 7.23. The performance evaluation of the C5.0 bio-kernel machine models constructed for six peptide datasets. (a) The models which employ the Dayhoff mutation matrix. (b) The models which employ the BLOSUM62 mutation matrix.

Table 7.5 shows the performance evaluation of AUC and MCC for the C5.0 bio-kernel models constructed for six peptide datasets and Figure 7.23 shows their ROC curves. It is as expected that the performance was not as good as previous bio-kernel models, for which the reason has been discussed when introducing the CART bio-kernel machine in the previous section of this chapter.

Both CART and C5.0 bio-kernel machines are unable to achieve good generalization performance, although they can deliver a model with explicit intelligence, i.e., a human-alike decision-making rule system, by which a hierarchical organization of partitioning rules makes an inference system convincing and explainable. The consideration of how to overcome the limitation of low generalization performance results in the development of another new intelligence exploitation algorithm, i.e., the random forest algorithm (Ho, 1998; Breiman, 2001; Hastie *et al.*, 2001).

First, let us analyze why a CART model and a C5.0 model are unable to achieve a good generalization performance. An example of a two-dimensional data space with two classes of data points is shown in Figure 7.24(a). Figure 7.24(b) shows a possible tree constructed by either CART or C5.0. As mentioned earlier, these algorithms search for the orthogonal partitioning rules as shown in Figure 7.24(b), i.e., the partitioning rules are always perpendicular to coordinates. Therefore, it is inevitable that a tree may be overgrown and inefficient. In most real-world applications, a data space may not be friendly to an algorithm which only searches for

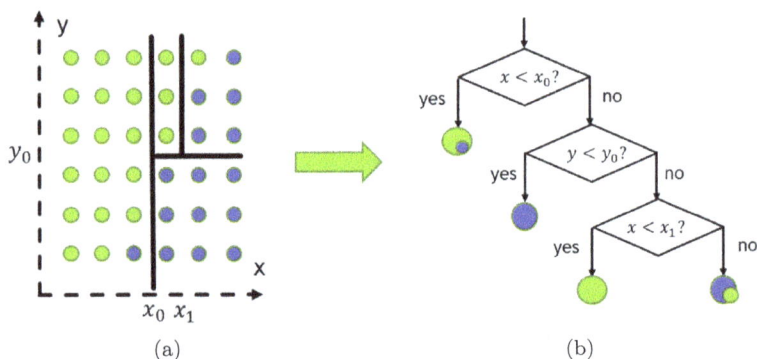

Figure 7.24. (a) A dataset for classification using the orthogonal partitioning rules. (b) The generated orthogonal decision tree. The leaf nodes with mixed color represent the fact that the resulting subspace is not pure for one class of data points. Adapted from Yang (2022a).

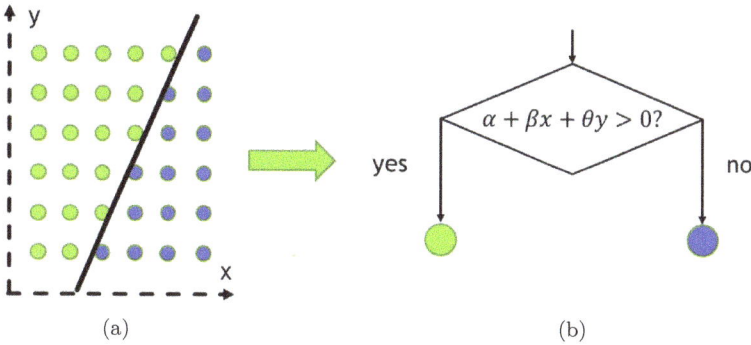

Figure 7.25. (a) A dataset for classification using an oblique partitioning rule. (b) The generated decision tree. Adapted from Yang (2022a).

the orthogonal partitioning rules. For instance, the pattern shown in Figure 7.24(a) has shown that it is more complex to an orthogonal decision tree constructed by an algorithm which partitions this space using orthogonal partitioning rules only.

Suppose we look at the problem from a different perspective, i.e., we design some oblique partitioning rules as shown in Figure 7.25. In this situation, things are different. A shown in Figure 7.25(a), having only one oblique partitioning rule may solve the problem, leading to a very efficient model. Figure 7.25(b) shows such a decision tree. It is definitely a nicer design and nicer model and this is the basic principle employed by the random forest algorithm (RF) (Ho, 1998; Breiman, 2001; Hastie *et al.*, 2001).

A more important feature of the random forest algorithm is that it can generate a pool (forest) of random trees. An optimization procedure is thus carried out among many random trees within the pool. Figure 7.26 shows one of the intelligent bio-kernel machine models constructed by RF for the O-linkage dataset.

Table 7.6 shows the performance evaluation (AUC and MCC) of the RF bio-kernel machine models constructed for six peptide datasets. Figure 7.27 shows their ROC curves. Compared with the intelligent bio-kernel machine models constructed by CART and C5.0, these RF bio-kernel machine models demonstrate much better performance.

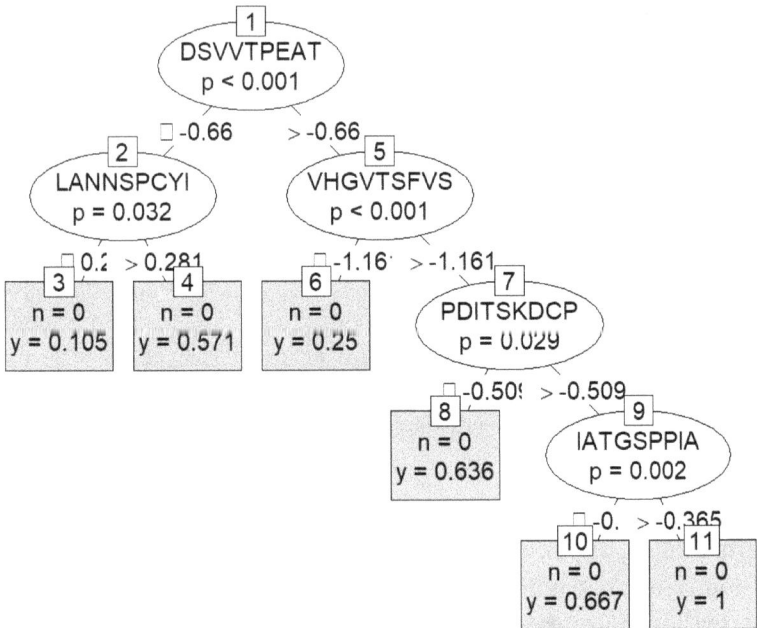

Figure 7.26. One of the RF bio-kernel machine models constructed for the O-linkage dataset.

Table 7.6. The performance of the RF bio-kernel models constructed for six peptide datasets. Dayhoff stands for the models constructed using the Dayhoff mutation matrix and BLOSUM62 stands for the models constructed using the BLOSUM62 mutation matrix.

Data name	Dayhoff		BLOSUM62	
	AUC	MCC	AUC	MCC
O-linkage	0.92	0.66	0.90	0.62
Factor Xa	0.94	0.58	0.93	0.64
Caspase	0.93	0.72	0.96	0.84
SARS	0.93	0.67	0.95	0.76
HCV	0.98	0.77	0.97	0.77
HIV	0.91	0.69	0.92	0.66

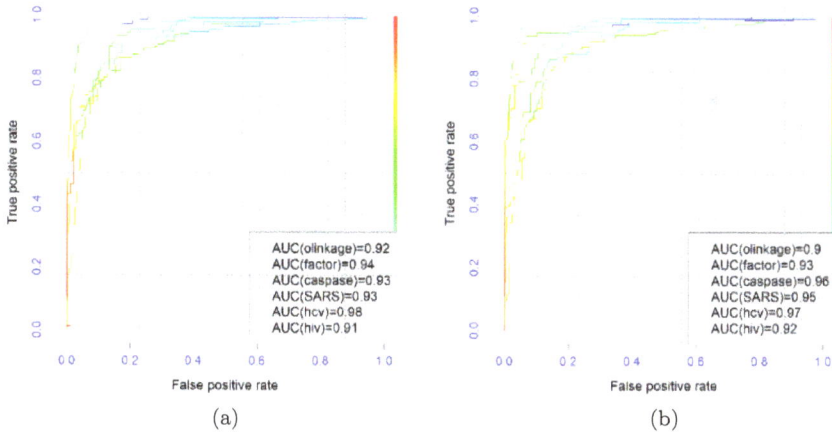

Figure 7.27. The ROC curves of the RF bio-kernel machine models constructed using six peptide datasets. (a) The models which employ the Dayhoff mutation matrix. (b) The models which employ the BLOSUM62 mutation matrix.

7.5 Deep bio-kernel machine

Deep learning or deep neural network has been widely researched recently in the machine learning community because of its enhanced generalization capability (Bengio *et al.*, 2015; Lecun *et al.*, 2015; Schmidhuber, 2015). Its basic concept is to increase the number of hidden layers in a conventional neural network model. For instance, Figure 7.28 shows such an example, where a network with one hidden layer is converted into a network with two hidden layers, resulting in a deep neural network.

In addition to the structure difference between a traditional neural network and a deep neural network, deep learning employs convolution to integrate feature extraction into one model process. In most real applications, feature extraction or feature selection is a tedious and time-consuming process. Although there have been many advanced approaches for feature extraction, especially feature selection, most of them are still ad hoc and problem specific (Bishop, 2006). However, a deep learning model or a deep network model pools all the features together into a learning process. Importantly, there

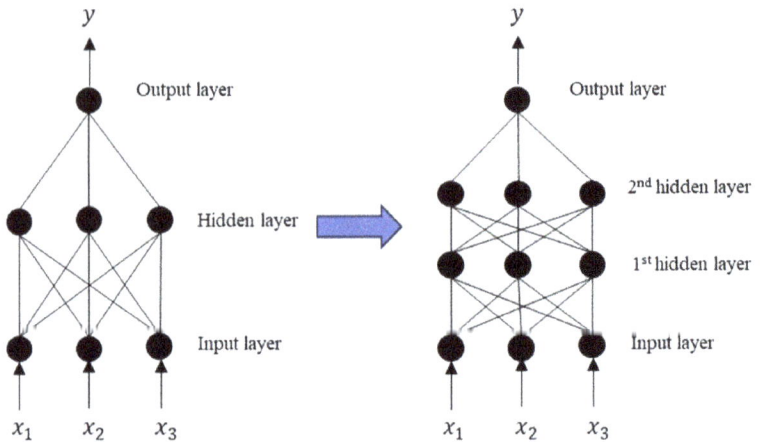

Figure 7.28. The conversion of a conventional neural network structure with one hidden neuron layer to a deep neural network structure with two hidden neuron layers.

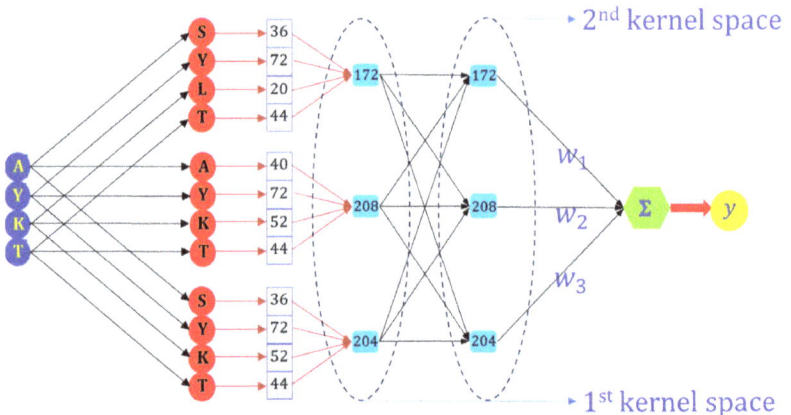

Figure 7.29. The illustration of a deep bio-kernel machine.

is no need for feature extraction as well as feature selection for some applications such as image recognition tasks.

As discussed in Chapter 3, the support vector machine (SVM) algorithm and the relevance vector machine (RVM) algorithm have very powerful capabilities to handle complex data with improved efficiency and accuracy. Figure 7.29 shows such a deep bio-kernel machine designed for peptide cleavage prediction (Yang and Chou, 2004a; Yang, 2022b).

In this network, there are two kernel space layers. The first kernel space is made by using a mutation matrix as seen in the LS bio-kernel machine or the FDA bio-kernel machine. Kernelization using a mutation matrix results in a kernel space $\mathbf{K} = (\hbar_{nh})_{n,h=1}^{N,H} \in \mathcal{R}^{N \times H}$ as shown in the following:

$$\mathcal{K}(\boldsymbol{v}_h, \mathbf{s}_n) = \sum_{r=1}^{R} \mathcal{m}(v_{hr}, s_{nr}) \qquad (7.84)$$

$\mathbf{K}$ is transferred to $\tilde{\mathbf{K}}$ by a mean–variance normalization process, i.e., $\mathbf{K} \rightarrow \tilde{\mathbf{K}}$. The second kernel space takes place in kernelization of $\tilde{\mathbf{K}}$:

$$\mathcal{K}(\tilde{\mathbf{K}}, \tilde{\mathbf{K}}) \rightarrow \tilde{\tilde{\mathbf{K}}} \qquad (7.85)$$

To construct an SVM bio-kernel model, one is required to adopt a grid search strategy to optimize an important hyper-parameter γ which controls the α parameters as seen in the following equation (refer to Chapter 3):

$$\max \left\{ \frac{1}{N} \sum_{n=1}^{N} \alpha_n - \frac{1}{2} \sum_{n=1}^{N} \sum_{m=1}^{N} y_n y_m \alpha_n \alpha_m \tilde{\hbar}_n^t \tilde{\hbar}_m \right\} \qquad (7.86)$$

Figure 7.30 shows the optimization of this hyper-parameter for the SVM bio-kernel models constructed for six peptide datasets. It can be seen that the model generalization performance can be optimized when the γ value is small.

Table 7.7 shows the performance (AUC and MCC) of these SVM bio-kernel models constructed for six peptide datasets. Figure 7.31 shows their ROC curves. It can be seen that the performance of these models has improved significantly compared with the LS bio-kernel models and the FDA bio-kernel models.

A constructed SVM bio-kernel model is capable of providing the support vector information. These support vectors are thus the kernel peptides which are on the boundary to discriminate between the functional and non-functional peptides. Figure 7.32 shows such

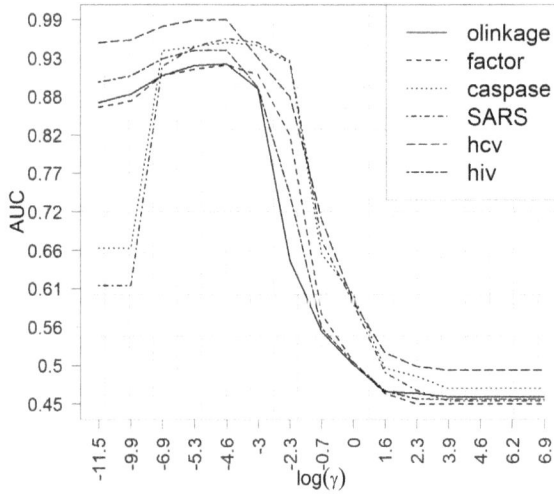

Figure 7.30. The optimization of the γ parameter for constructing the SVM bio-kernel models for six peptide datasets.

Table 7.7. The performance of the SVM bio-kernel models constructed for six peptide datasets. Dayhoff stands for the models constructed using the Dayhoff mutation matrix and BLOSUM62 stands for the models constructed using the BLOSUM62 mutation matrix.

	Dayhoff		BLOSUM62	
Data name	AUC	MCC	AUC	MCC
O-linkage	0.92	0.68	0.91	0.68
Factor Xa	0.92	0.64	0.95	0.7
Caspase	0.95	0.82	0.98	0.88
SARS	0.96	0.83	0.97	0.86
HCV	0.99	0.83	0.97	0.84
HIV	0.94	0.75	0.94	0.67

an example for the SARS dataset. The two-dimensional map was generated using Sammon mapping (Sammon, 1969). In this map, it can be seen that those peptides that are distributed within the boundaries are not selected as the support vectors and they are

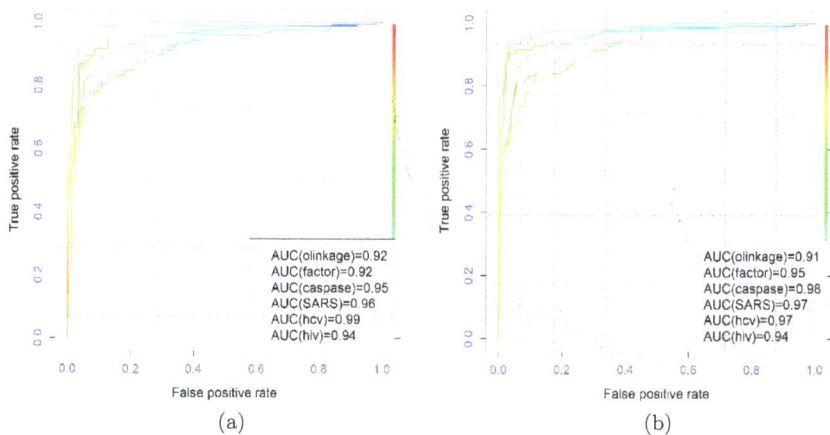

Figure 7.31. The ROC curves of the SVM bio-kernel models constructed for six peptide datasets. (a) The models which employ the Dayhoff mutation matrix. (b) The models which employ the BLOSUM62 mutation matrix.

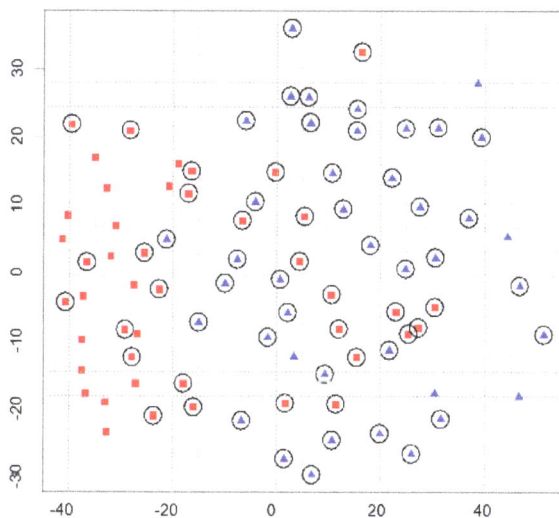

Figure 7.32. The Sammon map of the SVM bio-kernel model constructed for the SARS dataset. The filled triangles represent non-cleaved SARS peptides and the filled squares represent cleaved SARS peptides. The large open circles stand for the support vectors.

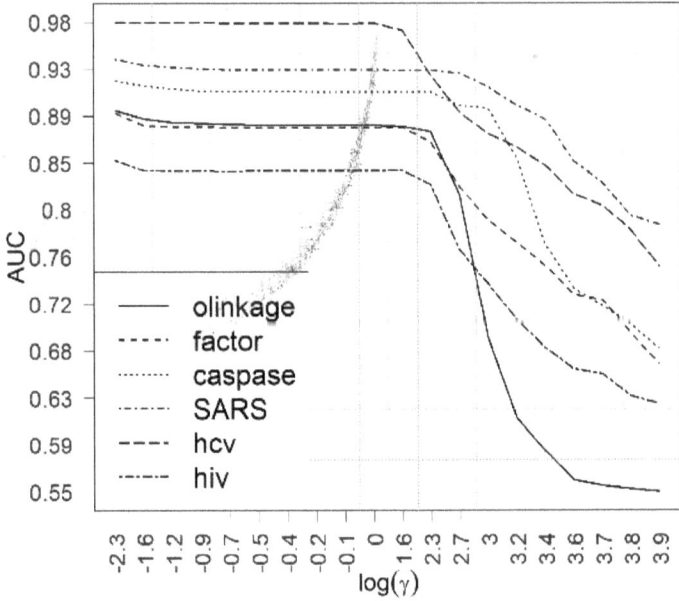

Figure 7.33. The optimization of the γ value for the RVM bio-kernel models constructed for six peptide datasets.

surrounded by a number of support vectors. For instance, the filled squares which are located on the left side of the map are those peptides.

To construct RVM bio-kernel models, the grid search approach is used to optimize the γ value. Figure 7.33 shows this optimization process for six peptide datasets. It also shows that the best γ value tends to be small.

Table 7.8 shows the performance evaluation (AUC and MCC) of the RVM bio-kernel models constructed for six peptide datasets. It can be seen that the performance has been improved significantly compared with the previously discussed LS bio-kernel models for these datasets. Figure 7.34 shows their ROC curves.

Table 7.8. The performance of the RVM bio-kernel models constructed for six peptide datasets. Dayhoff stands for the models constructed using the Dayhoff mutation matrix and BLOSUM62 stands for the models constructed using the BLOSUM62 mutation matrix.

Data name	Dayhoff		BLOSUM62	
	AUC	MCC	AUC	MCC
O-linkage	0.91	0.67	0.90	0.61
Factor Xa	0.94	0.62	0.95	0.68
Caspase	0.94	0.78	0.98	0.89
SARS	0.97	0.84	0.97	0.86
HCV	0.98	0.75	0.97	0.73
HIV	0.95	0.71	0.95	0.70

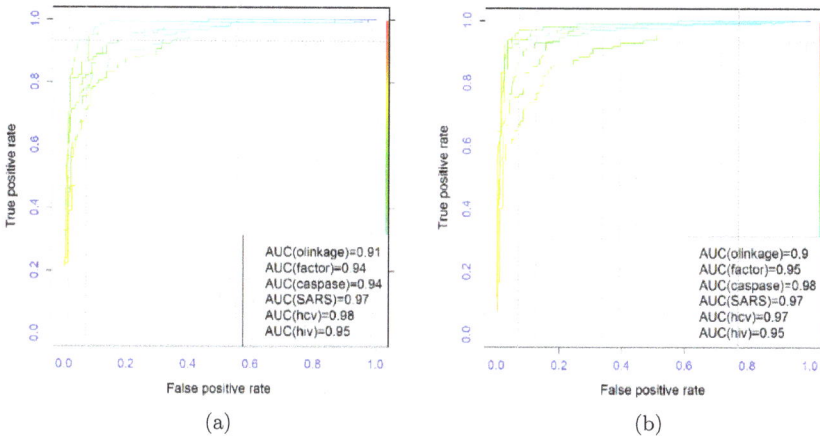

(a)   (b)

Figure 7.34. The ROC curves of the RVM bio-kernel models constructed for six peptide datasets. (a) The models which employ the Dayhoff mutation matrix. (b) The models which employ the BLOSUM62 mutation matrix.

7.6 Shrinkage bio-kernel machines versus deep bio-kernel machines

It must be noted that unlike shrinkage strategies such as the orthogonal bio-kernel machines, the deep bio-kernel machines have employed a different learning strategy. Their difference is the optimization target. The learning strategy used by the orthogonal bio-kernel machine is an iterative learning process, in which optimal variables are selected step by step till a predefined criterion is satisfied. However, the learning strategy employed by the deep bio-kernel machine is a batch learning approach, in which the optimal variables are selected in one learning process. Moreover, the optimal variables selected by algorithms which employ the shrinkage strategy, such as the orthogonal bio-kernel machine, are a subset of hypothetical kernel peptides. But, the optimal variables selected by deep bio-kernel machines are no longer hypothetical kernel peptides. Instead, the optimal variables selected by deep bio-kernel machines are a subset of training peptides.

In an orthogonal bio-kernel machine model, the selection of new orthogonal kernels must satisfy the following condition and the recruiting of new orthogonal kernels will terminate if this condition is violated:

$$\sum_{i=1}^{h} \overline{\overline{\mathfrak{F}}}(\mathbf{o}_i, \mathbf{y}) > 1 - \xi \qquad (7.87)$$

This equation implies the selection criterion as shown in the following:

$$\mathbf{O} = \{\operatorname*{argmax}_{h} \overline{\overline{\mathfrak{F}}}(\mathbf{o}_h, \mathbf{y})\} \qquad (7.88)$$

Figure 7.35 uses a simple and naïve example to illustrate how the shrinkage (Ridge, Lasso and orthogonal) bio-kernel machine carries out an optimization process in a kernel space. There are five candidate kernel peptides; hence, the original kernel space is a five-dimensional space. Because not all of these kernel peptides are required for the satisfactory performance of a bio-kernel machine model (based on the abovementioned Fisher test or AUC test), two

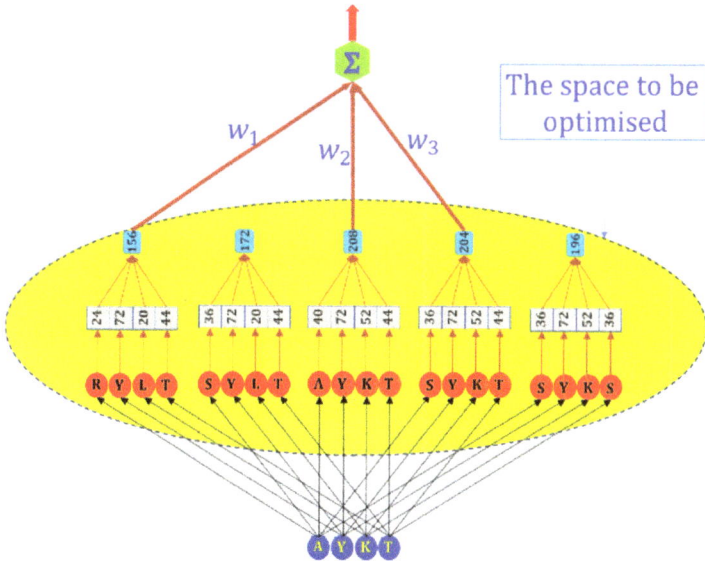

Figure 7.35. An illustration of how the shrinkage bio-kernel machine model works. The string AYKT stands for a training peptide. The strings RYLT, SYLT, AYKT, SYKT, and SYKS stand for five kernel peptides. The thick arrows stand for the connection between the output node (**y**) and the selected kernel peptides through a shrinkage learning process. The kernel peptides which have no connection with the output node (Σ) are the dropped kernels through a shrinkage optimization process.

of them are dropped. Only RYLT, AYKT, and SYKT are selected using an optimization process.

It can be seen that such a model output is a function of the kernels or orthogonal kernels in an orthogonal bio-kernel:

$$y(\mathbf{s}) = f(\langle \{\boldsymbol{v}_h\}_{h=1}^{H^-}, \mathbf{s}\rangle) \qquad (7.89)$$

This shows that a shrinkage bio-kernel model such as the orthogonal bio-kernel model optimizes a kernel set $(\{\boldsymbol{v}_h\}_{h-1}^{H^-})$, where $H^- < H$ stands for a reduced set of hypothetical bio-kernels employed in a final model.

But, the model output of a deep bio-kernel machine model is not a function of the hypothetical kernels. Instead, it has introduced a "secondary hidden space" like a deep neural network as mentioned earlier. Therefore, the output of a deep bio-kernel machine model is a function of a set of secondary kernels, which are actually the training peptides. For instance, the optimization function of the SVM bio-kernel model is thus defined as follows:

$$\max \left\{ \frac{1}{N} \sum_{i=1}^{N} \alpha_i - \frac{1}{2} \sum_{i=1}^{N} \sum_{j=1}^{N} y_i y_j \alpha_i \alpha_j \tilde{\boldsymbol{k}}_i^t \tilde{\boldsymbol{k}}_j \right\} \qquad (7.90)$$

Therefore, the optimization target of a deep bio-kernel model becomes the training peptide space (a subset of training peptides is selected in a final model) rather than the kernel peptide space. Figure 7.36 shows this principle.

Figure 7.37 further shows the difference between a shrinkage bio-kernel machine model and a deep bio-kernel machine model using a

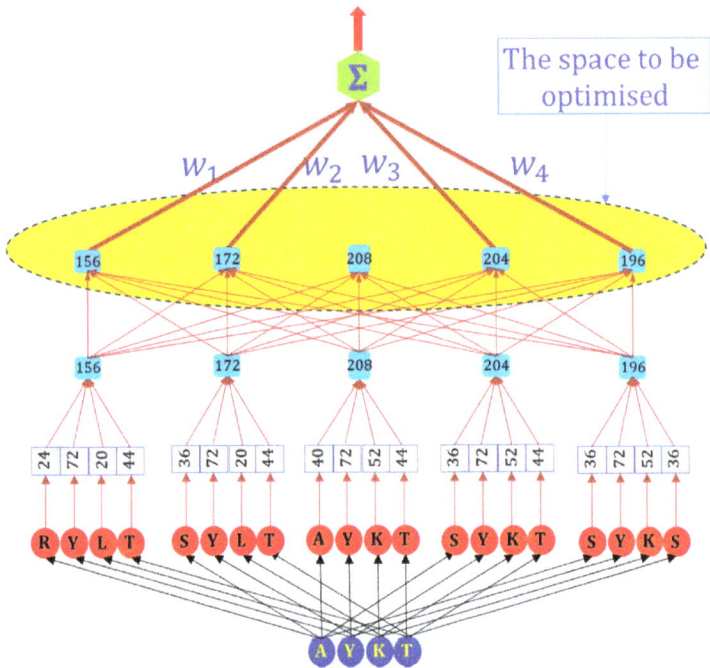

Figure 7.36.　An illustration of how a deep bio-kernel machine model works. The string AYKT stands a training or a testing peptide. The strings RYLT, SYLT, AYKT, SYKT, and SYKS stand for five kernel peptides. The thin arrows stand for the connection between the output node (**y**) and the selected secondary kernels (training peptides).

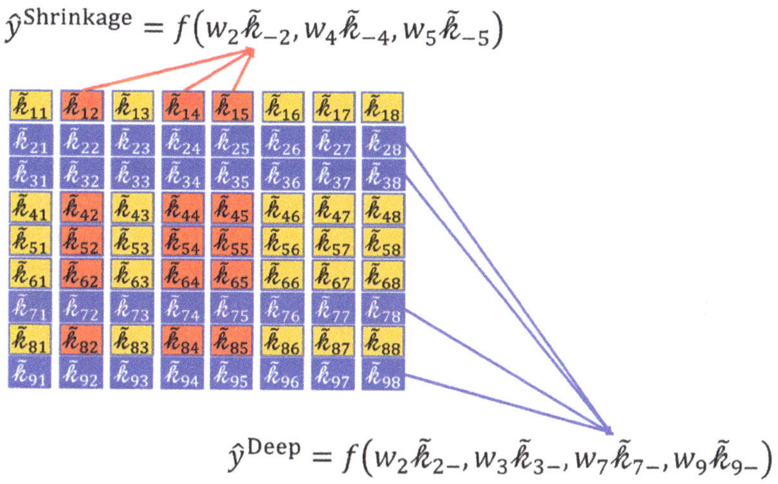

$$\hat{y}^{\text{Shrinkage}} = f\left(w_2 \tilde{k}_{-2}, w_4 \tilde{k}_{-4}, w_5 \tilde{k}_{-5}\right)$$

$$\hat{y}^{\text{Deep}} = f\left(w_2 \tilde{k}_{2-}, w_3 \tilde{k}_{3-}, w_7 \tilde{k}_{7-}, w_9 \tilde{k}_{9-}\right)$$

Figure 7.37. A comparison between a shrinkage bio-kernel machine model and a deep bio-kernel machine model.

simplified example where there are eight hypothetical kernel peptides and nine training peptides. Moreover, it is assumed that the same kernel space $\tilde{\mathbf{K}}$ is used for the illustration of both types of bio-kernel machines. The variables included in the finalized shrinkage bio-kernel machine model for a dataset are the column variables of a kernel matrix $\tilde{\mathbf{K}}$, such as $\tilde{k}_{-2}$, $\tilde{k}_{-4}$, and $\tilde{k}_{-5}$, where $\tilde{k}_{-h}$ stands for the hth column variable. However, the variables included in the finalized deep bio-kernel machine model are the row variables of a kernel matrix $\tilde{\mathbf{K}}$, such as $\tilde{k}_{2-}$, $\tilde{k}_{3-}$, $\tilde{k}_{7-}$, and $\tilde{k}_{9-}$, where $\tilde{k}_{n-}$ stands for the nth row variable.

7.7 Summary

This chapter has introduced several advanced bio-kernel machines. The first set includes the shrinkage bio-kernel models. The second set includes several intelligent bio-kernel machines. The third set includes the deep bio-kernel machines. Shrinkage bio-kernel machines and intelligent bio-kernel machines have a similar purpose, i.e., searching in a high-dimensional kernel space for an optimal set of hypothetical kernel peptides to generate a parsimonious bio-kernel machine model. But, the two approaches have different outcomes. The shrinkage bio-kernel machine model tends to maintain similar generalization capability by employing fewer kernel peptides. However, the intelligent bio-kernel machine model can deliver a set of partitioning rules for mimicking a human inference process, by which it is convenient to explain how a prediction is made and what the path is to approach a prediction. In contrast, the deep bio-kernel machine model will maintain the number of hypothetical kernel peptides but aim to reduce the number of training data points to be included in the finalized model. This is because the prediction of a novel data point using a deep bio-kernel model depends on a function of the training data points. To be specific, the prediction process using a deep bio-kernel machine model is a linear combination of the training data points. Therefore, reducing the number of training peptides in a deep bio-kernel machine model can improve the model generalization capability. Finally, the Bayesian bio-kernel machine aims to improve the model generalization capability by incorporating prior knowledge of the model parameters so that a robust model can be constructed using the constraint controlled by the prior probability during modeling.

Chapter 8

Fusion Bio-Kernel Models

There have been many amino acid mutation matrices, which
have been developed based on different protein sequences and
using different algorithms in the last 50 years. But, one question
remains unanswered: Which mutation matrix or which subset
of mutation matrices should be used for analyzing a specific
peptide dataset with a satisfactory generalization performance.
Because there is no *a priori* knowledge for optimal selection
of a mutation matrix for a specific peptide dataset, an ad hoc
grid search strategy may be the best strategy to employ though
it is inefficient. For instance, there will be 1023 combinations
(and hence 1023 modeling trials) for 10 mutation matrices.
Fusion technology, which is a popular approach used in machine
learning to deal with the uncertainty of utilizing heterogeneous
learning materials or resources, may be an appropriate approach
to consider due to the difficulty of mutation matrix selection
in constructing an integrative bio-kernel machine model. An
integrative bio-kernel machine model will fuse multiple mutation
matrices into one model for peptide pattern discovery and
analysis. This chapter thus starts the introduction of how
to apply the fusion technology for peptide pattern discovery
and analysis by employing heterogeneous mutation matrices.
Although the fusion technology may partially solve the problem,
whether we need to include all the available mutation matrices
is a question. Therefore, after introducing the fusion technology
for employing multiple mutation matrices for peptide pattern
discovery and analysis, a new approach is introduced in this
chapter. With this new approach, it is not required to fiddle
with mutation matrix selection to build an integrative bio-kernel
machine model. Instead, a new amino acid mutation matrix
is estimated through a learning process for a specific peptide

dataset. Such an amino acid mutation matrix is referred to as a data-driven mutation matrix. This chapter will show how it works and introduce two approaches for developing a data-driven mutation matrix for peptide pattern discovery and analysis based on a bio-kernel machine model.

8.1 Performance across amino acid mutation matrices

Since the development of the Dayhoff mutation matrix in 1978 (Dayhoff and Schwartz, 1978), many new mutation matrices have been developed based on different technologies as well as different protein sequence data. Table 8.1 shows 20 of the mutation matrices, which are available to use and hence are going to be investigated in this chapter, although there are many more alternatives.

Table 8.1. The 20 handy mutation matrices to be investigated in this chapter.

Name	Authors and the years of development
Azarya	Azarya-Sprinzak *et al.* (1997)
Benner	Benner *et al.* (1994)
Blake	Blake and Cohen (2001)
BLOSUM62	Henikoff and Henikoff (1992)
Crooks	Crooks *et al.* (2005)
Dayhoff	Dayhoff and Schwartz (1978)
Dosztanyi	Dosztanyi and Torda (2001)
Gonnet	Gaston *et al.* (1992)
Johnson	Johnson and Overington (1993)
Jones	Jones and Thornton (1992)
Kann	Kann *et al.* (2000)
Levin	Levin *et al.* (1986)
Mohana	Mohana (1987)
Muller	Muller (2001)
Naor	Naor *et al.* (1996)
Overington	Overington *et al.* (1990)
Prlic	Prlic *et al.* (2000)
Risler	Risler *et al.* (1988)
Russell	Russell *et al.* (1997)
Vogt	Vogt *et al.* (1995)

In almost all applications, there is an issue about how to utilize the heterogeneous materials and resources to understand the problem so as to model the collected data to generate a reliable and robust model for prediction and inference. Selecting the best data, such as a subset of features to describe or represent a problem efficiently, is non-trivial in machine learning. Utilizing existing knowledge or information to model a dataset efficiently is non-trivial, as well. Fusing heterogeneous data is thus a practical exercise in most real-world applications where machine learning approaches are employed to generate prediction and inference models (Kasturi and Acharya, 2005; Hert *et al.*, 2006; Jesneck *et al.*, 2006; Chen *et al.*, 2009; Ghanty and Pal, 2009; Metsis *et al.*, 2012; Christabel and Subhajini, 2023). Practically, to employ the fusion technology for a machine learning task, one needs to construct a unique model by integrating heterogeneous data into one input resource (such as an input matrix) for a learning process. Heterogeneous data have to be modeled separately if they cannot be modeled together in one model. Thus, a fusion process can take part in a learning process to generate a final decision-making system. A fusion model can be constructed in different ways. Among them, three of the most popular fusion technologies are described in the literature: data fusion technology (Hall and Llinas, 1997; Kasturi and Acharya, 2005; Ciuonzo *et al.*, 2013; Nadeem *et al.*, 2021), model fusion technology (Singh and Jagg, 2020), and decision-level fusion technology (Gumaei *et al.*, 2022; Badalassi *et al.*, 2023).

The data fusion technology (Nadeem *et al.*, 2021) associates and combines multiple (heterogeneous) data sources to construct a model for the purpose of improving the decision-making capability through the improvement of the data quality. From this, it is believed that fusion can enable a semi-automatic decision-making process during information processing. For instance, multiple genomic pieces of data can be combined into a clustering algorithm for cluster analysis of genes (Kasturi and Acharya, 2005). The other example is the integration and alignment of multiple sources of sensors into one detection system (Hall and Llinas, 1997; Ciuonzo *et al.*, 2013).

When adopting the model fusion technology (Singh and Jagg, 2020), inter-structure intervention between several models ensure that there is much information to be communicated over the models for knowledge sharing. For instance, neurons from different neural networks were communicated or aligned to generate a fused neural network as the final decision-making model (Singh and Jagg, 2020).

The decision-level fusion technology usually involves a process of combining predictions from multiple models to generate unified predictions (Badalassi *et al.*, 2023). For instance, a soft voting technique was used to generate final decisions regarding how to monitor patients' health after COVID-19 for efficient treatment (Gumaei *et al.*, 2022). The voting was applied to outputs from multiple models, such as the random forest (Ho, 1998; Breiman, 2001; Hastie *et al.*, 2001), gradient boosting (Piryonesi and El-Diraby, 2020), and extreme gradient boosting algorithms (Sagi and Rokach, 2021).

If heterogeneous mutation matrices are available, a bio-kernel machine model may employ all of them, a part of them, or a single one of them. The selection of mutation matrix is thus the key to complete a peptide pattern discovery and analysis task. The performance variation of the bio-kernel machine models constructed for six peptide datasets is investigated. The investigation includes two aspects. The first is how model performance varies across datasets. The second is how the model performance varies across heterogeneous mutation matrices.

Figure 8.1(a) shows the model performance (AUC) variation across six peptide datasets. It can be seen that the LS bio-kernel models constructed for the O-linkage peptide dataset have the largest variation when employing 20 mutation matrices. The standard deviation of AUC for this dataset is 0.076. The LS bio-kernel models constructed for the HCV peptide dataset have the smallest variation when employing 20 mutation matrices. The standard deviation of AUC for this dataset is 0.006, which is more than one magnitude lower than that of the O-linkage dataset. This analysis shows that

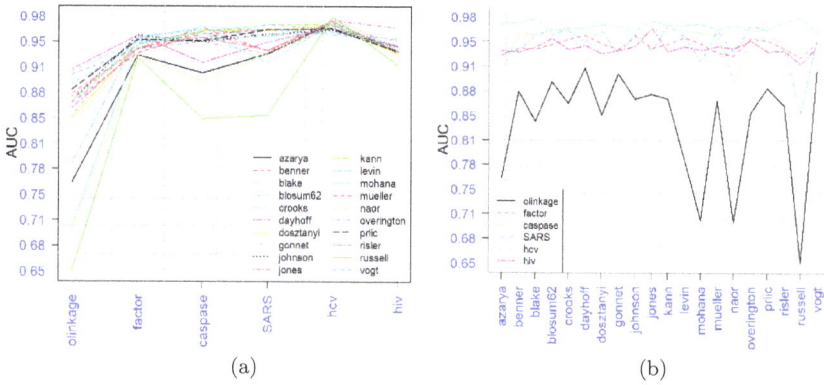

Figure 8.1. (a) The AUC variation of the bio-kernel machine models. (b) The AUC variation across mutation matrices. The models were constructed using the LS bio-kernel machine.

some datasets are sensitive to mutation matrices, while others are not. The bio-kernel machine models constructed for the O-linkage dataset are very sensitive to the use of mutation matrices. The bio-kernel machine models constructed for the HCV dataset are insensitive to the use of mutation matrices. Figure 8.1(b) shows how AUC varies across the 20 mutation matrices. It can be seen that the Russell mutation matrix demonstrated the maximum variation for six peptide datasets. The standard deviation of AUC for this mutation matrix is 0.117. But, the Vogt mutation has the smallest variation for six peptide datasets. The standard deviation of AUC for this mutation matrix is 0.019, which is about one magnitude lower than that of the Russell mutation matrix.

Figure 8.2 further shows the error bar chart adapted from Figure 8.1(a). This investigation shows that it is difficult to know in advance how large the performance deviation will be prior to a tedious modeling/analysis process. Therefore, some advanced technologies should be considered to overcome this uncertainty if we do need to select the best mutation matrices for constructing a bio-kernel machine model for a peptide dataset.

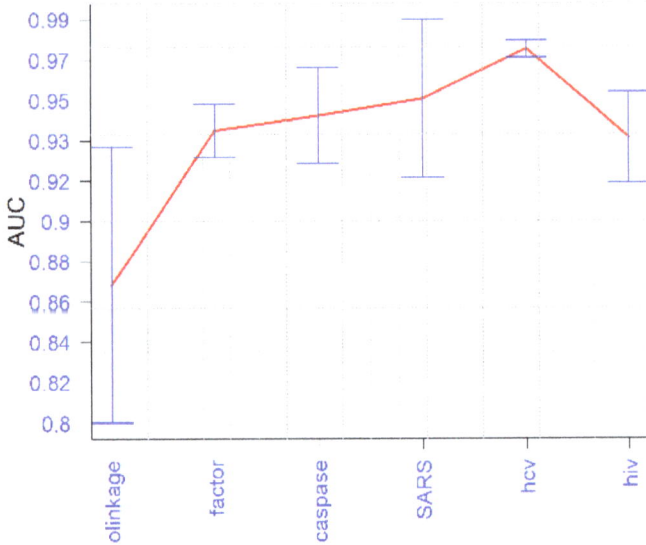

Figure 8.2. The error bar (one standard deviation) chart of the AUC performance of the LS bio-kernel models constructed using 20 mutation matrices for six peptide datasets.

8.2   Fusion across amino acid mutation matrices

In the previous section, the performance variation of different mutation matrices to construct a bio-kernel machine model was discussed. In this section, fusion technology is employed to deal with the uncertainty of selecting a proper mutation matrix for modeling and analyzing a specific peptide dataset. The first technology used is the so-called the data fusion technology.

Applying the data fusion technology to a bio-kernel machine model requires a data organization process prior to a modeling process. The employment of the data fusion technology in this chapter adopts the following transformation:

$$\mathbf{K}^F = \frac{1}{M} \sum_{m=1}^{M} \rho(\mathbf{K}_m) \qquad (8.1)$$

where M stands for the number of mutation matrices, $\mathbf{K}_m$ (or $\tilde{\mathbf{K}}_m$) is a kernel matrix generated using the mth mutation matrix, $\mathbf{K}^F$ (or $\tilde{\mathbf{K}}_F$) is the fusion kernel matrix, and ρ is the sigmoid function. The

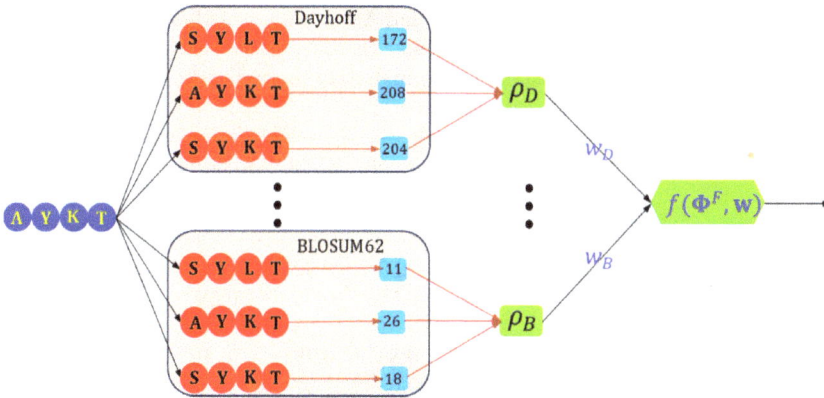

Figure 8.3. The data fusion structure for the bio-kernel machine. In the diagram, two blocks are shown, which are the Dayhoff block and the BLOSUM62 block. In these two blocks, the non-gap pairwise alignment scores between the training peptide and three kernel peptides using two mutation matrices are shown. For instance, the alignment scores between the training peptide and the first kernel peptide (SYLT) are 172 and 11 using the Dayhoff mutation matrix and the BLOSUM62 mutation matrix, respectively. The other 18 blocks are omitted for simplicity.

fusion output based on the fusion kernel matrix $\mathbf{K}^F$ is defined as follows, where $\mathbf{w}$ is the model parameter set:

$$\hat{\mathbf{y}} = f(\mathbf{K}^F, \mathbf{w}) \tag{8.2}$$

Figure 8.3 shows the data fusion structure of the bio-kernel machine, where a training peptide AYKT is inputted to the system to align with three hypothetical kernel peptides (SYLT, AYKT, and SYKT) using 20 mutation matrices as listed in Table 8.1. These 20 kernel matrices $(\mathbf{K}_1, \mathbf{K}_2, \ldots, \mathbf{K}_{20})$ are fused into one fusion kernel matrix $\mathbf{K}^F$ through a summed sigmoid function transformation. The fused kernel matrix $\mathbf{K}^F$ is used to build a predictive bio-kernel model $f(\mathbf{K}^F, \mathbf{w})$. Note that this predicted output vector $\hat{\mathbf{y}}$ is used for decision-making or model performance evaluation through a comparison against the known class label vector $\mathbf{y}$ of the training peptides.

After the fusing kernel matrix has been generated, a bio-kernel machine model is constructed. Table 8.2 shows the performance

Table 8.2. The AUC and MCC values of the
LS bio-kernel models and the deep bio-kernel
models constructed through data fusion for six
peptide datasets.

	LS bio-kernel		SVM bio-kernel	
Data name	AUC	MCC	AUC	MCC
O-linkage	0.90	0.59	0.92	0.66
Factor Xa	0.94	0.55	0.95	0.68
Caspase	0.95	0.81	0.98	0.88
SARS	0.98	0.83	0.98	0.89
HCV	0.97	0.65	0.98	0.82
HIV	0.95	0.69	0.96	0.76

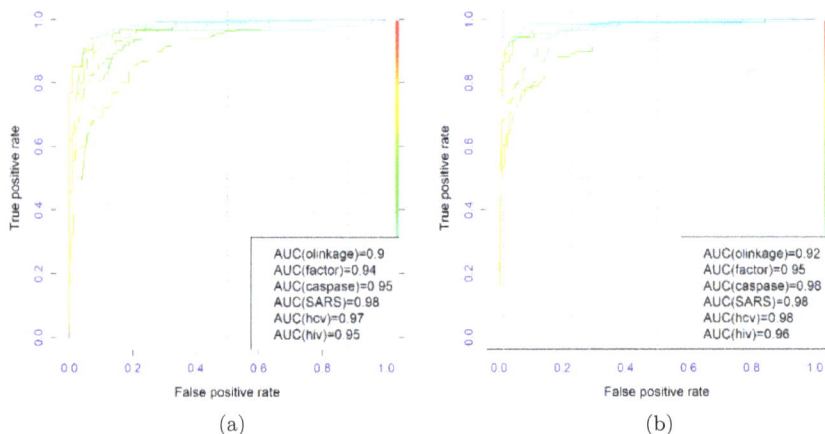

Figure 8.4. The ROC curves of the data-fused bio-kernel machine models
constructed for six peptide datasets. (a) The LS bio-kernel models. (b) The SVM
bio-kernel models.

(AUC and MCC) of the LS bio-kernel models and deep bio-kernel machine (SVM bio-kernel) models constructed through data fusion for six peptide datasets. Figure 8.4 shows their ROC curves. Compared with Figures 8.1 and 8.2, it can be seen that the uncertainty of these bio-kernel machine models has been removed when using the data fusion technology. Moreover, the performance

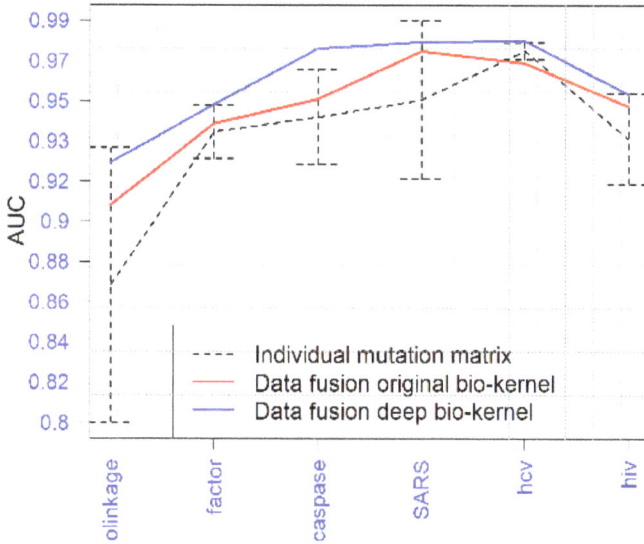

Figure 8.5. The performance comparison between the data-fused bio-kernel machine models and the bio-kernel machine models without fusion. The thick and solid lines represent the performance of the data-fused bio-kernel machine models. "Original" and "deep" stand for the LS and the SVM bio-kernel models, respectively.

of the data-fused deep bio-kernel models is also slightly better than the performance of the deep bio-kernel models without fusion.

Figure 8.5 shows the AUC comparison between the bio-kernel models constructed using each of 20 mutation matrices and the data-fused bio-kernel models. It can be seen that both data-fused bio-kernel machine models outperformed the bio-kernel models without data fusion.

The model fusion technology applied to the bio-kernel machine models, which employ multiple heterogeneous mutation matrices, employs the following procedure. First, one bio-kernel model (inner model) is constructed for each kernel matrix $\mathbf{K}_m$ (or $\tilde{\mathbf{K}}_m$) generated using one mutation matrix, where $\mathbf{w}_m$ is the model parameter set and $\mathbf{q}_m$ is the output of the mth model:

$$\mathbf{q}_m = f_m(\mathbf{K}_m, \mathbf{w}_m) \to \mathbf{y} \qquad (8.3)$$

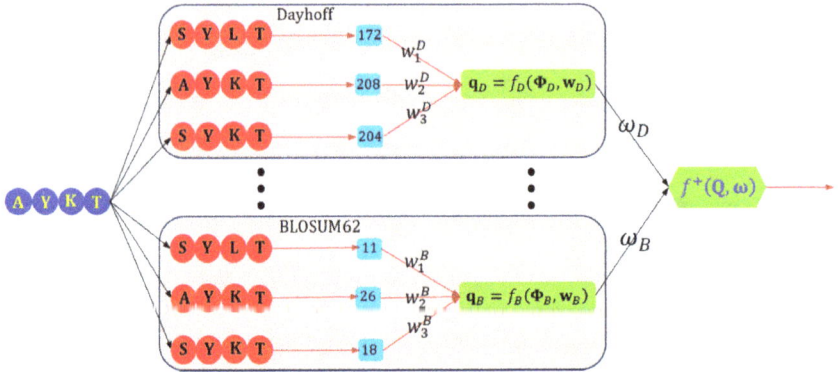

Figure 8.6. The model fusion structure for the bio-kernel machine. In the diagram, two inner models are shown, which are the Dayhoff inner model and the BLOSUM62 inner model. The other 18 inner models are omitted for simplicity.

Collecting all the inner bio-kernel machine models constructed using all the mutation matrices leads to a super "kernel" matrix $\mathbf{Q}$:

$$\mathbf{Q} = (\mathbf{q}_m)_{m=1}^{M} \qquad (8.4)$$

A second layer model (fusing model f^+) is constructed for this super kernel matrix $\mathbf{Q}$ as shown in the following, where $\boldsymbol{\omega}$ is a set of parameters employed by the fusing model:

$$\mathbf{y} = f^+(\mathbf{Q}, \boldsymbol{\omega}) \qquad (8.5)$$

Figure 8.6 shows the structure of the model fusion technology applied to the bio-kernel machine. In this diagram, there are two layers of modeling. The lower layer is composed of the bio-kernel machine models (inner models). Training peptides are inputted to the inner models. A training process is carried out to generate inner models. For instance, the model parameters shown in Figure 8.6 can be $\mathbf{w}^D = (w_1^D, w_2^D, w_3^D)$ for employing three kernel peptides (SYLT, AYKT, and SYKT) in the Dayhoff model as well as $\mathbf{w}^B = (w_1^B, w_2^B, w_3^B)$ for the same three kernel peptides in the BLOSUM62 inner model. Finally, the fusion model is trained again with the model parameters $\boldsymbol{\omega} = (\omega_D, \ldots, \omega_B)$, where ω_D and ω_B stand for the weights associated with the Dayhoff and BLOSUM62 inner models, respectively.

Table 8.3. The AUC and MCC values for the fused LS bio-kernel models and the fused deep bio-kernel models constructed through model fusion for six peptide datasets.

Data name	LS bio-kernel		SVM bio-kernel	
	AUC	MCC	AUC	MCC
O-linkage	0.92	0.66	0.92	0.68
Factor Xa	0.96	0.64	0.96	0.76
Caspase	0.97	0.91	0.98	0.89
SARS	0.99	0.89	0.98	0.92
HCV	0.98	0.62	0.99	0.86
HIV	0.96	0.73	0.96	0.82

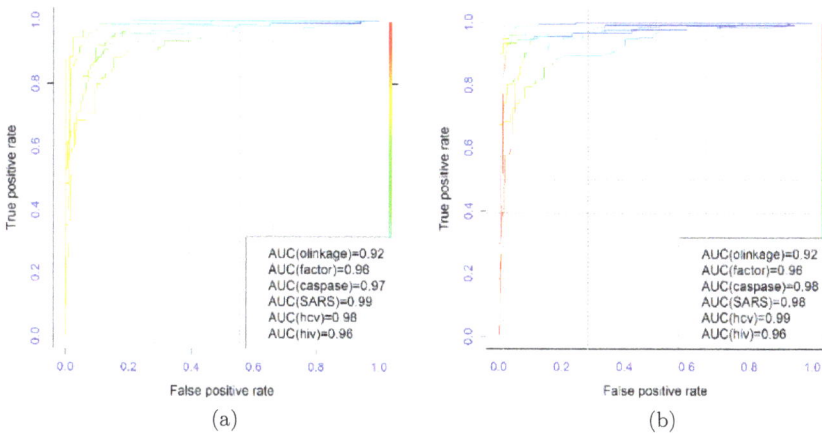

Figure 8.7. The ROC curves of the bio-kernel machine models constructed using the model fusion technology for six peptide datasets. (a) The LS bio-kernel models. (b) The SVM bio-kernel models.

Table 8.3 shows the performance (AUC and MCC) for the bio-kernel models constructed through model fusion for six peptide datasets. Figure 8.7 shows their ROC curves. It can be seen that the performance has slightly improved. For instance, the AUC values have been increased to 0.99 for the fused LS bio-kernel model constructed for the SARS dataset and the fused SVM bio-kernel model constructed for HCV dataset. Figure 8.8 shows the performance

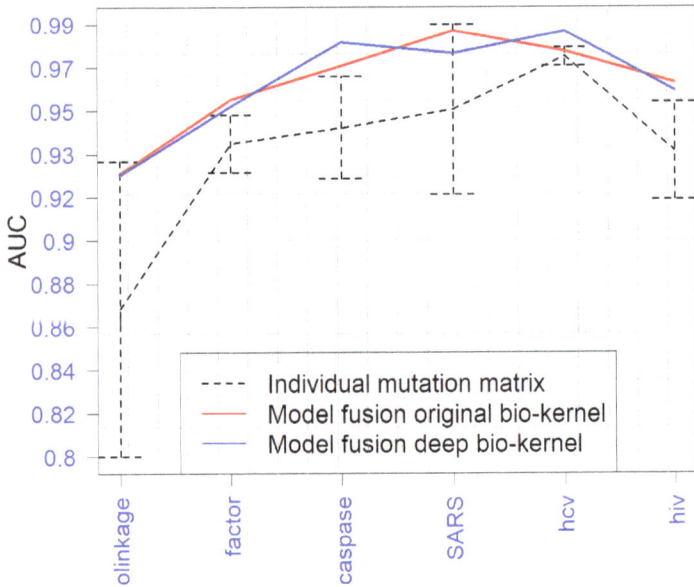

Figure 8.8. The performance comparison between the fused bio-kernel machine models and the bio-kernel machine models constructed without fusion. The thick and solid lines represent the fused bio-kernel machine models. "Original" and "deep" stand for the LS and the SVM bio-kernel models, respectively.

comparison between the fused bio-kernel machine models and the bio-kernel machine models constructed without model fusion. It can be seen again that the model fusion technology indeed improved the performance.

The decision-level fusion technology does not need to train an extra model for the relationship between the outputs of the models constructed based on different mutation matrices and the class labels of the training peptides. Instead, the decision-level fusion tends to establish a probabilistic decision-making agent based on the bio-kernel models constructed before fusion. The decision-level fusion technology applied to the bio-kernel machine thus employs the following procedure. First, a density function or a conditional probability is estimated for each bio-kernel machine model which employs one mutation matrix:

$$p(q_{nm}^{\text{Tr}}|y_n = k) = \frac{1}{\sqrt{2\pi\sigma_{km}^2}}\exp\left(-\frac{(q_{nm}^{\text{Tr}} - \mu_{km})^2}{2\sigma_{km}^2}\right) \tag{8.6}$$

where q_{nm}^{Tr} is the prediction of the mth model for the nth training peptide, μ_{km} is the mean of q_{nm}^{Tr} over the class k, and σ_{km}^2 is the variance. Note that $k \in \{1, 2\}$ and $K = 2$ for this scenario. The likelihood of q_{nm}^{Tr} is defined as follows, where ω_{km} is called a mixing coefficient:

$$p\left(q_{nm}^{\text{Tr}}\right) = \sum_{k=1}^{K} \omega_{km} p\left(q_{nm}^{\text{Tr}}|y_n = k\right) \tag{8.7}$$

The posterior probability that a peptide belongs to the kth class is defined as follows:

$$p(k|q_{nm}^{\text{Tr}}) = \frac{\omega_{km} p(q_{nm}^{\text{Tr}}|y_n = k)}{\sum_{g=1}^{K} \omega_{gm} p(q_{nm}^{\text{Tr}}|y_n = g)} \tag{8.8}$$

The likelihood function of the mth model is defined as follows:

$$\mathcal{L}_m = \prod_{n=1}^{N} p\left(q_{nm}^{\text{Tr}}\right) \tag{8.9}$$

In this model, μ_{km}, σ_{km}^2, and ω_{km} are model parameters which are estimated using the likelihood maximization approach (Bishop, 1996; Duda et al., 2000; Webb, 2002). Through maximizing the likelihood defined previously, three parameters (μ_{km}, σ_{km}^2, and ω_{km}) can be estimated using the Expectation Maximization procedure (Dempster et al., 1977; Meng and van Dyk, 1997). The estimation procedure of the three parameters is shown in the following, along with the estimation of μ_{km}:

$$\hat{\mu}_{km} = \frac{\sum_{n=1}^{N} p(y_n = k|q_{nm}^{\text{Tr}})q_{nm}^{\text{Tr}}}{\sum_{n=1}^{N} p(y_n = k|q_{nm}^{\text{Tr}})} = \frac{\sum_{n=1}^{N} p(k|q_{nm}^{\text{Tr}})q_{nm}^{\text{Tr}}}{\sum_{n=1}^{N} p(k|q_{nm}^{\text{Tr}})} \tag{8.10}$$

Note that the summation of the posterior probabilities across classes is always one:

$$\sum_{k=1}^{K} p(k|q_{nm}^{\text{Tr}}) \equiv 1 \qquad (8.11)$$

The estimation of σ_{km}^2 is shown as follows:

$$\hat{\sigma}_{km}^2 = \frac{1}{K} \frac{\sum_{n=1}^{N} p(k|q_{nm}^{\text{Tr}})(q_{nm}^{\text{Tr}} - \mu_{km})^2}{\sum_{n=1}^{N} p(k|q_{nm}^{\text{Tr}})} \qquad (8.12)$$

Finally, the mixing coefficients are estimated as follows:

$$\hat{\omega}_{km} = \frac{1}{N} \sum_{n=1}^{N} p(k|q_{nm}^{\text{Tr}}) \qquad (8.13)$$

After the parameter set $(\mu_{km}, \sigma_{km}^2,$ and $\omega_{km})$ has been estimated based on training datasets, the estimation on testing dataset can be done for each model as shown in the following:

$$p(k|q_{nm}^{\text{Te}}) = \frac{\omega_{km} p(q_{nm}^{\text{Te}}|y_n = k)}{\sum_{g=1}^{K} \omega_{gm} p(q_{nm}^{\text{Te}}|y_n = g)} \qquad (8.14)$$

The next step, which is important, is to assemble the predictions from M models, which are constructed using the M mutation matrices. This is the task of making an ensemble of M predictions for each testing peptide, i.e., the decision-level fusion:

$$\{p(k|q_{n1}^{\text{Te}}), p(k|q_{n2}^{\text{Te}}), \ldots, p(k|q_{nM}^{\text{Te}})\} \qquad (8.15)$$

What it is needed is such a weighted prediction for this ensemble:

$$\hat{y}_n = \sum_{m=1}^{M} \pi_m p(k|q_{nm}^{\text{Te}}) \qquad (8.16)$$

It must be noted that π_m is independent of the construction of the probabilistic model $p(k|q_{nm}^{\text{Tr}})$. The value of π_m is defined as follows to account for the performance of each bio-kernel model constructed based on one mutation matrix, where ϑ_m is the AUC value estimated on the training dataset in the mth bio-kernel machine model:

$$\pi_m = \frac{\vartheta_m}{\sum_{i=1}^{M} \vartheta_i} \qquad (8.17)$$

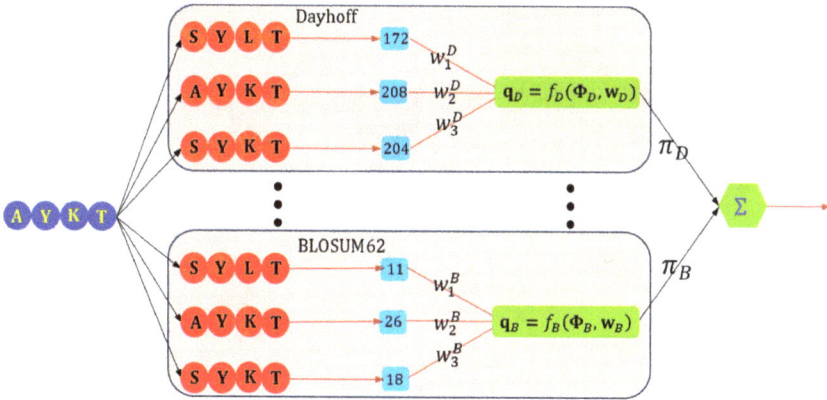

Figure 8.9. The decision-level fusion structure for the bio-kernel machine. In the diagram, two inner models are shown, which are the Dayhoff inner model and the BLOSUM62 inner model. The other 18 inner models are omitted for simplicity.

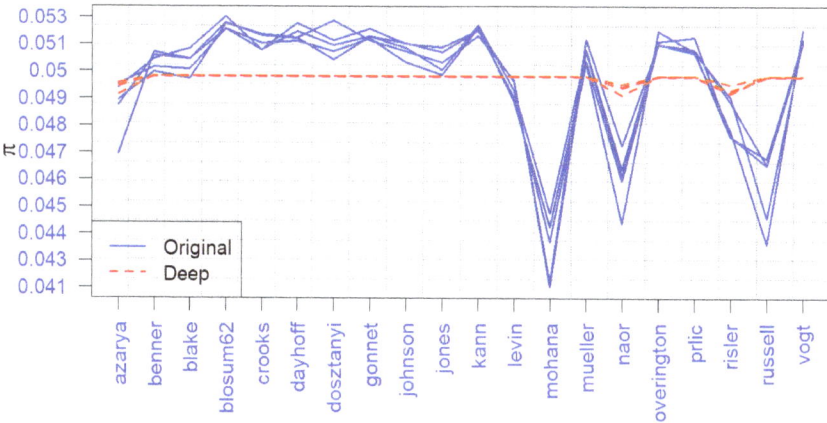

Figure 8.10. The weight pattern of the decision-level fused bio-kernel models generated using the AUC across 20 mutation matrices for the O-linkage dataset. "Original" and "Deep" stand for the decision-level fused LS and SVM bio-kernel models, respectively.

Figure 8.9 shows the structure of decision-level fusion employed for the bio-kernel machine in this chapter.

Figure 8.10 shows the distribution of the weights (π) of 20 mutation matrices used for modeling the O-linkage dataset using the

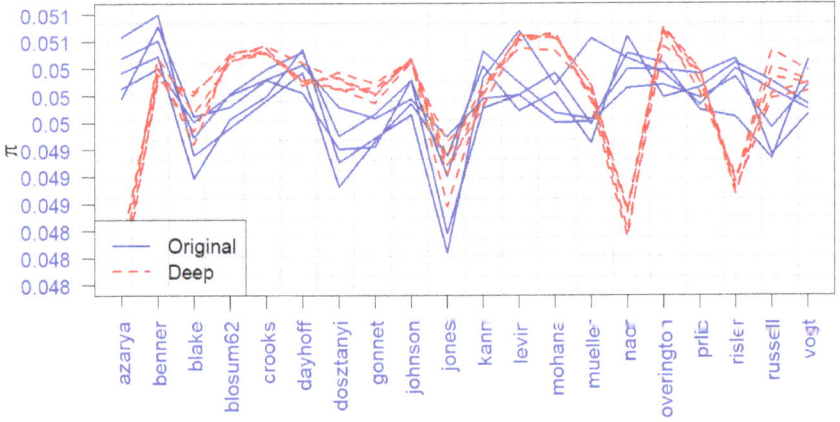

Figure 8.11. The weight pattern of the decision-level fused bio-kernel models generated using the AUC across 20 mutation matrices for the Caspase dataset. "Original" and "Deep" stand for the decision-level fused LS and SVM bio-kernel models, respectively.

decision-level fusing bio-kernel machine model. There are five sets because 5-fold cross-validation was used. The figure shows that the LS bio-kernel models are sensitive to the employment of different mutation matrices, and hence the weights change dramatically. However, the weights do not change much in the deep bio-kernel machine (SVM bio-kernel). Figure 8.11 shows another scenario, where a comparison is made by analyzing the weight distribution for the Caspase dataset. It can be seen that both the LS bio-kernel models and the SVM bio-kernel models were sensitive to the use of different mutation matrices. This indicates that the fusion of different mutation matrices for different datasets requires a careful investigation.

Figure 8.12 shows the heatmap of the posterior probabilities generated from the bio-kernel machine models constructed for the O-linkage dataset when employing different mutation matrices. It can be seen that the posterior probabilities of the LS bio-kernel models are generally clustered into two parts (Figure 8.12(a)), one part corresponds to the small weights group (such as the bio-kernel machine models constructed using the Mohana, Naro, and Russel mutation matrices) shown in Figure 8.10, while the other part

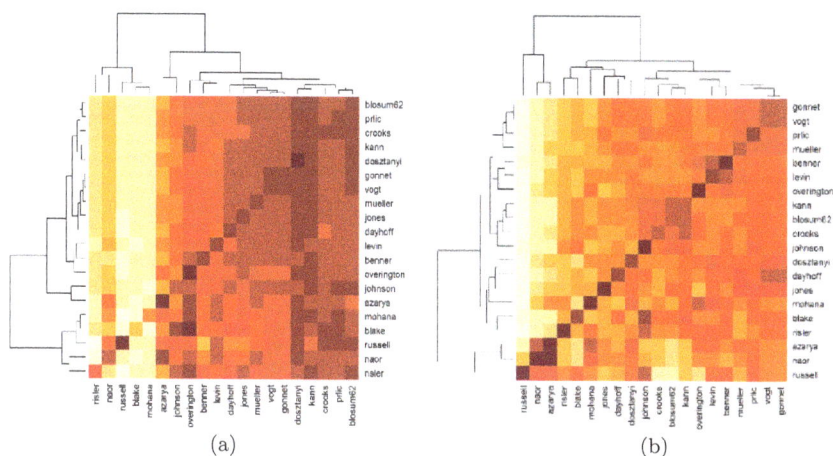

(a)   (b)

Figure 8.12. The pattern of the correlation between posterior probabilities of 20 bio-kernel machine models using 20 different mutation matrices. The heatmap was generated for the O-linkage dataset. (a) The LS bio-kernel model. (b) The SVM bio-kernel model.

Table 8.4. The AUC and MCC values for the decision-level fused LS and deep bio-kernel machine models constructed for six peptide datasets.

Data name	LS bio-kernel		SVM bio-kernel	
	AUC	MCC	AUC	MCC
O-linkage	0.91	0.61	0.93	0.69
Factor Xa	0.96	0.68	0.96	0.69
Caspase	0.97	0.87	0.97	0.86
SARS	0.98	0.85	0.99	0.88
HCV	0.98	0.76	0.99	0.79
HIV	0.95	0.75	0.95	0.79

corresponds to the great weights group shown in Figure 8.10. But, Figure 8.12(b) shows a small correlation between the posterior probabilities among the bio-kernel machine models constructed using different mutation matrices.

Table 8.4 shows the performance (AUC and MCC) of the decision-level fused bio-kernel models. Figure 8.13 shows the ROC of these

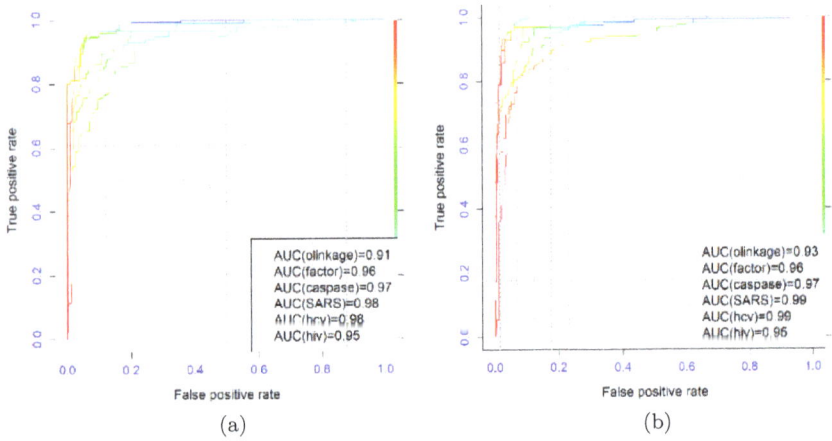

Figure 8.13. The ROC curves of the decision-level fused bio-kernel machine models constructed for six peptide datasets. (a) The LS bio-kernel models. (b) The SVM bio-kernel models.

models. The performance of the decision-level fused SVM bio-kernel models seem to be the best so far.

8.3   Data-driven estimated mutation matrix for bio-kernel machine

Applying the fusion technology to the bio-kernel machine can help deal with the uncertainty of mutation matrix selection. However, it still requires work in modeling, including computing memory and computing speed. This section introduces an approach by which we do not have the uncertainty of selecting the best mutation matrix for a dataset and also do not need to fuse multiple mutation matrices. Instead, an amino acid mutation matrix is directly estimated for the given peptide dataset that is under investigation.

Chapter 5 has shown that the amino acid composition pattern is extremely conserved at certain residues in cleaved or post-translational modified peptides. Due to this property, a data-driven mutation matrix can be estimated based on the functional peptides of a dataset for employing a machine learning algorithm such as the bio-kernel machine to model and analyze peptide data. Two

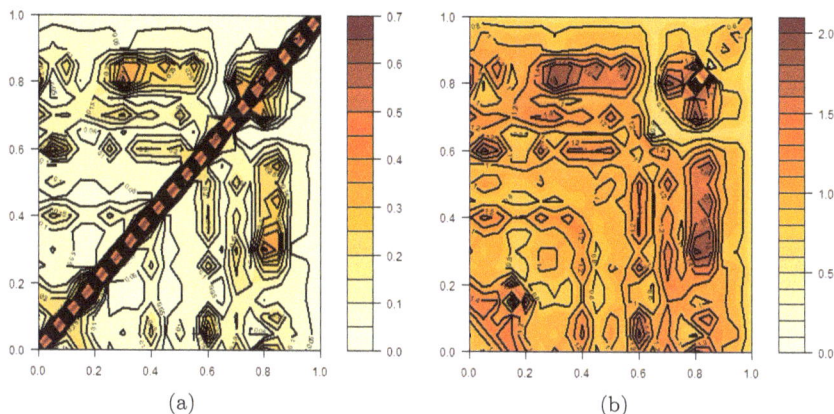

(a)   (b)

Figure 8.14. The contours for the PAM mutation matrices estimated for the Factor Xa dataset using the Dayhoff approach. (a) The λ value is one. (b) The λ value is 10.

approaches are considered for this practice. The first is the data-driven approach for estimating a mutation matrix, which is based on the Dayhoff approach introduced in Chapter 5. The second is Monte Carlo simulation, which generates many candidate mutation matrices by drawing samples randomly. The final estimated mutation matrix is the mean of these candidate mutation matrices.

Figure 8.14 shows the contours of the PAM mutation matrices estimated based on two λ values ($\lambda = 1$ and $\lambda = 10$) for the Factor Xa dataset. The comparison shows that when the λ value is small, the diagonal entries of the estimated mutation matrix are significantly greater than the off-diagonal entries of the mutation matrix. But, when the λ value increases, the values of the diagonal entries do not show dominance over the values of the off-diagonal entries. To search for the optimal λ value for each of six peptide datasets, 10 λ values have been tested and the resulting models have been evaluated using the AUC value. Figure 8.15 shows the variation of the AUC values within the LS bio-kernel models and the deep bio-kernel models constructed based on the mutation matrices estimated using the data-driven approach for six peptide datasets. It can be seen that the deep bio-kernel models demonstrate a smaller variation

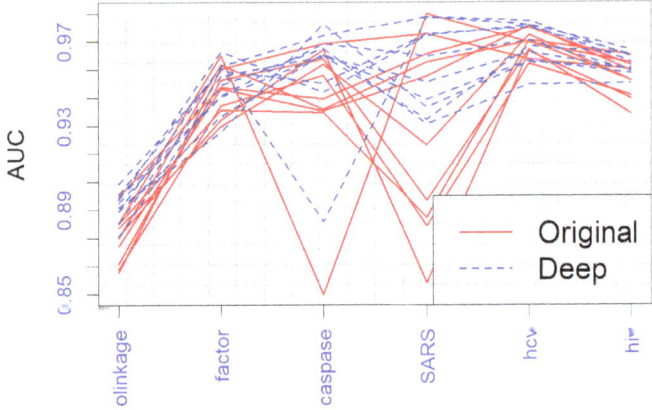

Figure 8.15. The AUC variation of the bio-kernel machine models constructed based on the data-driven estimated mutation matrices using different λ values for six peptide datasets. "Original" and "deep" stand for the LS and the deep bio-kernel models, respectively.

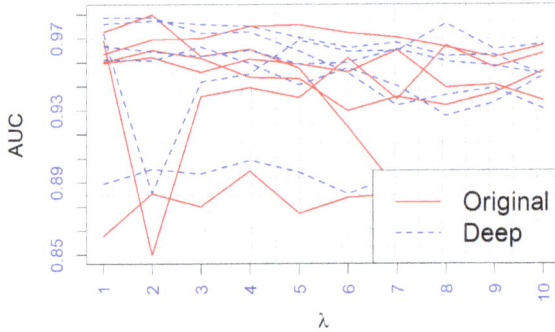

Figure 8.16. The AUC variation of the bio-kernel machine models constructed based on the data-driven estimated mutation matrices using different λ values. "Original" and "deep" stand for the LS and the deep bio-kernel models, respectively.

in performance. Moreover, the deep bio-kernel models generally demonstrate a better performance than the LS bio-kernel models.

Figure 8.16 shows the AUC variation across the λ values. It means that there is no single optimal λ value for all datasets. But, generally speaking, the deep bio-kernel machine models tend to employ smaller λ values with slightly better performance.

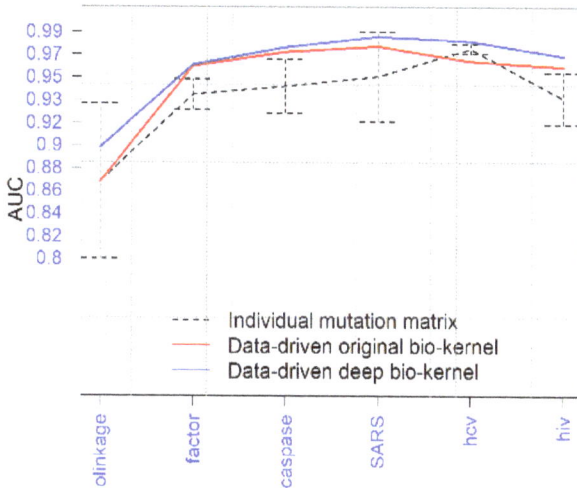

Figure 8.17. The comparison between the bio-kernel models constructed based on the data-driven estimated mutation matrices for six peptide datasets and the bio-kernel machine models constructed using each of the 20 mutation matrices. The thick and solid lines represent the LS bio-kernel models and the deep bio-kernel models, respectively. "Original" and "deep" stand for the LS and the deep bio-kernel models, respectively.

Figure 8.17 shows the comparison between the bio-kernel models constructed for six peptide datasets based on the data-driven estimated mutation matrices and the bio-kernel machine models constructed using each of the 20 published mutation matrices. It can be seen that the two sets of models have comparable performance.

Table 8.5 shows the performance (AUC and MCC) of the LS bio-kernel models and the deep bio-kernel models constructed using the data-driven estimated mutation matrices when the λ value is two. Figure 8.18 shows their ROC curves. It can be seen that the performance of these bio-kernel machine models is comparable to the performance of the bio-kernel machine models constructed using the Dayhoff or BLOSUM62 mutation matrices described in Chapter 6 and Chapter 7. This is an important property so far and also proves that estimating a mutation matrix directly for a specific dataset under investigation and using such an estimated mutation matrix for peptide pattern discovery and analysis is a feasible approach.

Table 8.5. The AUC and MCC values for the LS bio-kernel models and the deep bio-kernel models constructed using the data-driven estimated mutation matrices (PAM2) for six peptide datasets.

Data name	LS bio-kernel		SVM bio-kernel	
	AUC	MCC	AUC	MCC
O-linkage	0.87	0.58	0.90	0.65
Factor Xa	0.96	0.62	0.96	0.73
Caspase	0.97	0.01	0.97	0.91
SARS	0.98	0.81	0.99	0.85
HCV	0.97	0.65	0.98	0.85
HIV	0.96	0.75	0.97	0.77

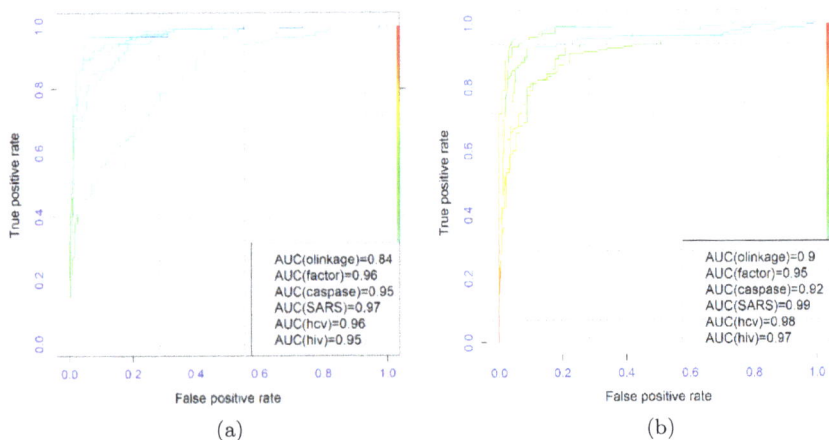

Figure 8.18. The ROC curves of the bio-kernel models constructed for six peptide datasets using the data-driven estimated mutation matrices (PAM2). (a) The LS bio-kernel model. (b) The SVM bio-kernel model.

There have been many different approaches for estimating mutation matrices other than the Dayhoff approach, including the maximum likelihood and Bayesian approaches (Muller *et al.*, 2002; Tseng and Liang, 2006; Trudgian and Yang, 2007; Huelsenbeck *et al.*, 2008). Most of them need to estimate a phylogenetic tree for whole

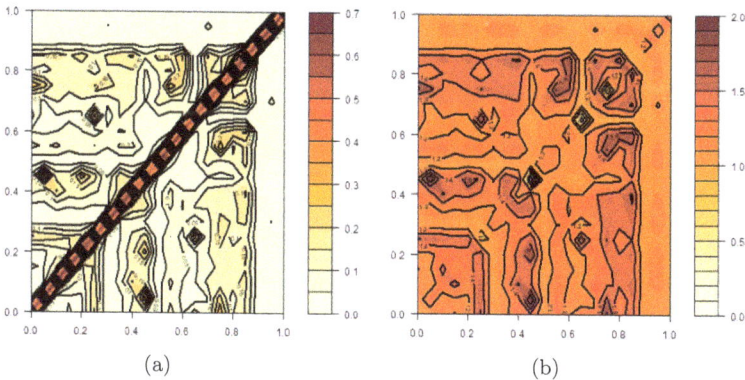

(a) (b)

Figure 8.19. The mutation matrices estimated using the Monte Carlo approach for the SARS dataset. (a) PAM1 ($\lambda = 1$). (b) PAM10 ($\lambda = 10$).

sequences before estimating a mutation matrix. The Monte Carlo approach is used here to estimate a mutation matrix for a dataset. A dataset of functional peptides is used to draw 100 samples. For each draw, a candidate mutation matrix is estimated. The final estimated mutation matrix is the mean of all these 100 candidate mutation matrices. Figure 8.19 shows the estimated mutation matrices for the SARS dataset using the Monte Carlo approach. One has the λ value set to one and the other has the λ value set to 10. It can be seen that they show the same pattern as seen in Figure 8.14. When the λ value is small, the diagonal entry values dominate the matrix; when the λ value is large, this dominance pattern thus disappears.

Figure 8.20 shows the comparison between the bio-kernel machine models constructed based on the estimated mutation matrices using the Monte Carlo approach and those models constructed using each of the 20 published mutation matrices. The chart shows that the former set of the bio-kernel machine models slightly outperforms the latter set of the bio-kernel machine models. Table 8.6 shows the performance (AUC and MCC) of the LS bio-kernel models and the SVM bio-kernel models constructed using the mutation matrices estimated by the Monte Carlo approach for six peptide datasets. Figure 8.21 shows their ROC curves. It can be seen that

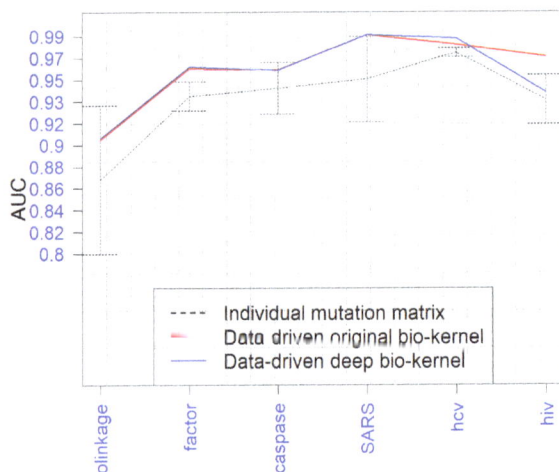

Figure 8.20. The comparison between the bio-kernel models constructed based on the mutation matrices estimated using the Monte Carlo approach and the bio-kernel machine models constructed using each of the 20 mutation matrices for six peptide datasets. The thick and solid lines represent the LS bio-kernel models and the SVM bio-kernel models, respectively. "Original" and "deep" stand for the LS and the SVM bio-kernel models, respectively.

Table 8.6. The AUC and MCC values for the LS bio-kernel models and the SVM bio-kernel models constructed using the Monte Carlo simulation of estimating data-driven mutation matrices (PAM3) for six peptide datasets.

Data name	LS bio-kernel		SVM bio-kernel	
	AUC	MCC	AUC	MCC
O-linkage	0.91	0.65	0.91	0.66
Factor Xa	0.94	0.60	0.94	0.78
Caspase	0.95	0.91	0.96	0.86
SARS	0.97	0.85	0.98	0.85
HCV	0.97	0.59	0.99	0.82
HIV	0.97	0.79	0.96	0.82

the performance has been improved, especially for the LS bio-kernel models based on the Monte Carlo estimation approach compared with the LS bio-kernel models constructed using the mutation matrix estimated based on a single dataset.

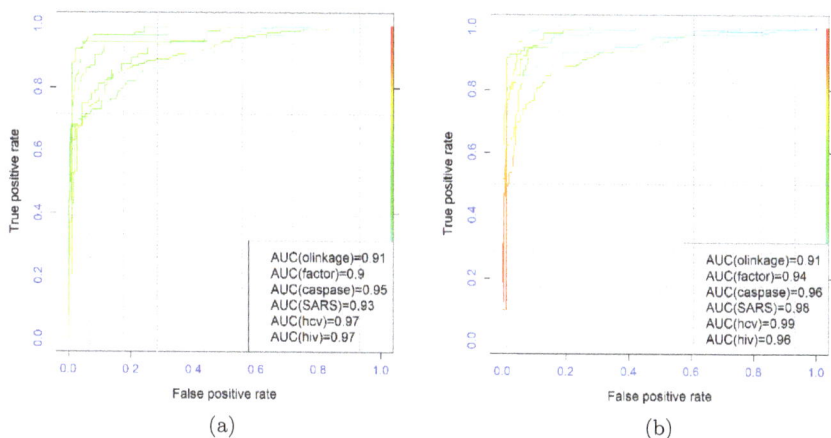

Figure 8.21. The ROC curves of the bio-kernel machine models constructed using the mutation matrices estimated based on the Monte Carlo approach for six peptide datasets. (a) The LS bio-kernel models. (b) The SVM bio-kernel models.

8.4 Summary

This chapter has introduced two new strategies regarding the handling of the uncertainty of employing a single mutation matrix when constructing a bio-kernel machine model for peptide pattern discovery and analysis. There are too many mutation matrices available and more and more mutation matrices are under development. Moreover, this study has shown that the model performance varies when using different mutation matrices. In some circumstance, this variation is significant. Having understood that there is a great degree of deviation between mutation matrices and between models constructed using different mutation matrices, thus the question is how to handle different mutation matrices when one is required to analyze a peptide dataset using the bio-kernel machine.

The first strategy is the use of the fusion technology. Three handy and commonly used fusion technologies in machine learning have been employed in this chapter for handling the uncertainty issue when employing a single mutation matrix to construct a bio-kernel model for a dataset. Although only a small subset of mutation matrices (20 of them) have been used, the result is promising.

The second strategy introduced in this chapter is that rather than employing fusion technology to avoid the mutation matrix selection uncertainty, a mutation matrix is directly estimated from a peptide dataset which is under investigation. This kind of mutation matrix estimated from a given peptide dataset is referred to as the data-driven estimated mutation matrix. The Dayhoff approach and the Monte Carlo approach have been used for the data-driven estimation of a mutation matrix for a peptide dataset. The approaches have demonstrated that these models have comparable generalization performance. Therefore, like whole-sequence analysis, peptide pattern discovery and analysis can also rely on the filed-oriented estimation of a mutation matrix to make the task easy to run.

Chapter 9

Visualize Bio-Kernel Machines

Visualization is one of the main research areas in machine learning. Its power in data structure investigation has made it a popular research subject in real-world applications, including biological data modeling and analysis. Having understood the discriminant power of bio-kernel machines in previous chapters, it is interesting to examine how a bio-kernel machine model can be visualized using traditional visualization algorithms and what discrimination power a visualized bio-kernel machine model has. This chapter therefore introduces visualization approaches for bio-kernel machines. Although there are many visualization approaches in the literature, two of them are introduced in this chapter due to their solid statistical background as well as the flexibility in use. The first is principal component analysis and the second is the self-organizing map. The former is a linear approach, while the latter is a nonlinear approach. Combined with bio-kernel machines, they are referred to as bio-kernelized visualizers in this book. This chapter will first show how these visualization approaches are bio-kernelized and then investigate the discrimination power a bio-kernelized visualizer may have for peptide data.

9.1 Bio-kernelized principal component analysis

As an extension of principal component analysis (PCA) in the kernel space, the kernelized principal component analysis (kPCA) approach has been researched over the decades (Scholkopf et $al.$, 1998; Huang et $al.$, 2009). Suppose a data (input) space denoted by $\mathbf{X} = (\mathbf{x}_1, \mathbf{x}_2, \ldots, \mathbf{x}_N)$ has been mapped to a feature space $\mathbf{\Phi} = (\phi_1(\mathbf{x}_1), \phi_2(\mathbf{x}_2), \ldots, \phi_N(\mathbf{x}_N))$ using a feature extraction approach.

Whether such a feature space is measurable or not is not important because the final working space is not this feature space. Instead, a kernel space transformed from this feature space is the final working space for kPCA. The transformation from a feature space to a kernel space is implemented using a kernel function as mentioned earlier:

$$\mathbf{K} = \mathcal{K}(\boldsymbol{\Phi}, \boldsymbol{\Phi}) = \langle \boldsymbol{\Phi}, \boldsymbol{\Phi} \rangle \tag{9.1}$$

The kernel space $\mathbf{K}$ is a numerical space. Suppose the space is case-wise centralized:

$$\boldsymbol{\mu}_n = \frac{1}{N} \sum_{n=1}^{N} \hbar_n = 0 \tag{9.2}$$

The covariance matrix can thus be simply defined as follows:

$$\boldsymbol{\Sigma} = \frac{1}{N} \sum_{n=1}^{N} \hbar_n \hbar_n^t \tag{9.3}$$

Diagonalizing $\boldsymbol{\Sigma}$ leads to

$$\lambda \mathbf{v} = \boldsymbol{\Sigma} \mathbf{v} \tag{9.4}$$

where λ is an eigenvalue and $\mathbf{v}$ is an eigenvector for the diagonalized model. A simple example is used to illustrate how it works. Suppose the covariance matrix of a very simple kernel space is shown as follows:

$$\boldsymbol{\Sigma} = \begin{pmatrix} 6 & 4 \\ 10 & 9 \end{pmatrix} \tag{9.5}$$

A linear transformation can be applied to this space so as to map this space to a one-dimensional space. To do this, a vector is required, for instance,

$$\mathbf{v} = \begin{pmatrix} 1 \\ 2 \end{pmatrix} \tag{9.6}$$

The transformation can be done using the following calculation:

$$\boldsymbol{\Sigma} \times \mathbf{v} = \begin{pmatrix} 6 & 4 \\ 10 & 9 \end{pmatrix} \times \begin{pmatrix} 1 \\ 2 \end{pmatrix} = \begin{pmatrix} 14 \\ 28 \end{pmatrix} = 14 \times \begin{pmatrix} 1 \\ 2 \end{pmatrix} = 14 \times \mathbf{v} = \lambda \mathbf{v} \tag{9.7}$$

Note that $\mathbf{v} = (1, 2)^t$ is an eigenvector and $\lambda = 14$ is an eigenvalue. The abovementioned mapping represents the following relationship:

$$\text{cov matrix} \times \text{eigen vector} = \text{eigen value} \times \text{eigen vector} \quad (9.8)$$

PCA is designed in a way to find these eigenvectors, which are referred to as principal components (PCs), which are also treated as the coordinates in a mapping space. The number of PCs always equals the number of variables, i.e., the column dimension of a kernel matrix. PCA is a parametric approach because it works by finding a linear mapping (projection) function so that

$$\mathbf{Z} = \Sigma \times \mathbf{V} \quad (9.9)$$

where $\mathbf{Z}$ is the mapping space and $\mathbf{V}$ is a mapping matrix of the collection of eigenvectors. With a proper learning process, Eq. (9.9) can be written as follows:

$$\Sigma \times \mathbf{V} = \Sigma \times \mathbf{v}_1 + \Sigma \times \mathbf{v}_2 + \cdots + \Sigma \times \mathbf{v}_H \quad (9.10)$$

where H is the column dimension of a kernel matrix $\mathbf{K}$. Each column vector of $\mathbf{V}$ ($\mathbf{v}_h$), is an eigenvector as mentioned earlier. This equation can be further rewritten as follows:

$$\Sigma \times \mathbf{V} = \lambda_1 \times \mathbf{v}_1 + \lambda_2 \times \mathbf{v}_2 + \cdots + \lambda_H \times \mathbf{v}_H \quad (9.11)$$

where λ_h is the hth eigenvalue corresponding to the hth eigenvector $\mathbf{v}_h$. In PCA learning, we will ensure the following condition for mutual orthogonality between eigenvectors in $\mathbf{V}$:

$$\mathbf{V}^t\mathbf{V} = \mathbf{I} \quad (9.12)$$

It is equivalent to the following equation:

$$\mathbf{v}_h \cdot \mathbf{v}_h = 1 \quad (9.13)$$

This orthogonality ensures that the transpose of $\mathbf{V}$ equals the inverse of $\mathbf{V}$ according to the matrix operation:

$$\mathbf{V}^t = \mathbf{V}^{-1} \quad (9.14)$$

Based on this, we have the following important rule, in which the variance in the mapping space ($\mathbf{Z}$) equals the variance in the covariance matrix of the kernel space ($\mathbf{K}$):

$$\mathbf{ZZ}^t = \mathbf{\Sigma VV}^t\mathbf{\Sigma}^t = \mathbf{\Sigma\Sigma}^t \tag{9.15}$$

Moreover, the total variance of $\mathbf{Z}$ equals the sum of the variances of all coordinates (principal components) because the coordinates of $\mathbf{Z}$ are mutually orthogonal, i.e.,

$$\mathrm{var}(\mathbf{Z}) = \sum_{h=1}^{H} \mathrm{var}(\mathbf{z}_h) \tag{9.16}$$

Another important property of PCA is that the variances of PCs have a descending order across the eigenvectors shown in the following:

$$\mathrm{var}(\mathbf{z}_1) > \mathrm{var}(\mathbf{z}_2) > \cdots > \mathrm{var}(\mathbf{z}_H) \tag{9.17}$$

Figure 9.1 shows an example in two-dimensional space, where it can be seen that the first PC (PC$_1$) catches the greater variance of

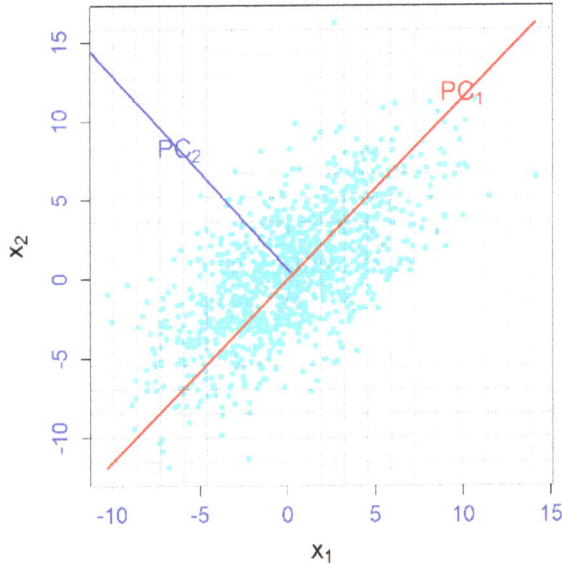

Figure 9.1. An illustration of how PCs are laid out in a space. The first PC (PC$_1$) lies in the direction of the greater variance of the data and the second PC (PC$_2$) lies in the direction which is perpendicular to the first PC to represent the rest (smaller) of the variance of the data.

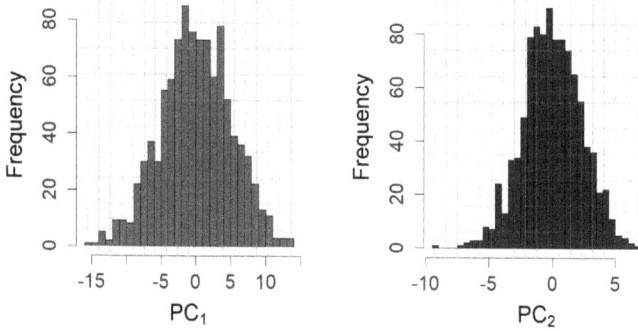

Figure 9.2. The densities of two PCs of the PCA model constructed for the data shown in Figure'9.1.

the data, while the second PC (PC_2), which is perpendicular to the first PC, catches the rest (smaller) of the variance of the data. This is why Eq. (9.17) holds true.

Figure 9.2 shows the density of two PCs. It can be seen that PC_1 has a greater variance compared to PC_2.

The abovementioned properties show that the employment of the first few principal components (PCs) for modeling will not lose much power of pattern discovery due to the null correlation between PCs and their descending variances across eigenvectors. But, this will not happen in the original kernel space, in which variables or kernels are highly mutually correlated. The comparison shown in the following proves this.

Figure 9.3 shows the correlation pattern between the hypothetical kernel peptides for the SARS peptide dataset (Yang, 2005a). It can be seen that there is a large quantity of mutual correlation between the hypothetical kernel peptides that are treated as variables in a model. In particular, the top-right block of the heatmap displays an interesting property. In that block, the dominance of the variances of the kernel peptides has diminished compared to the bottom-left block, where the variances of the kernel peptides are much greater than the covariances between the kernel peptides. This means that although the kernel peptides clustered into the bottom-left block demonstrate low mutual correlation, the kernel peptides clustered

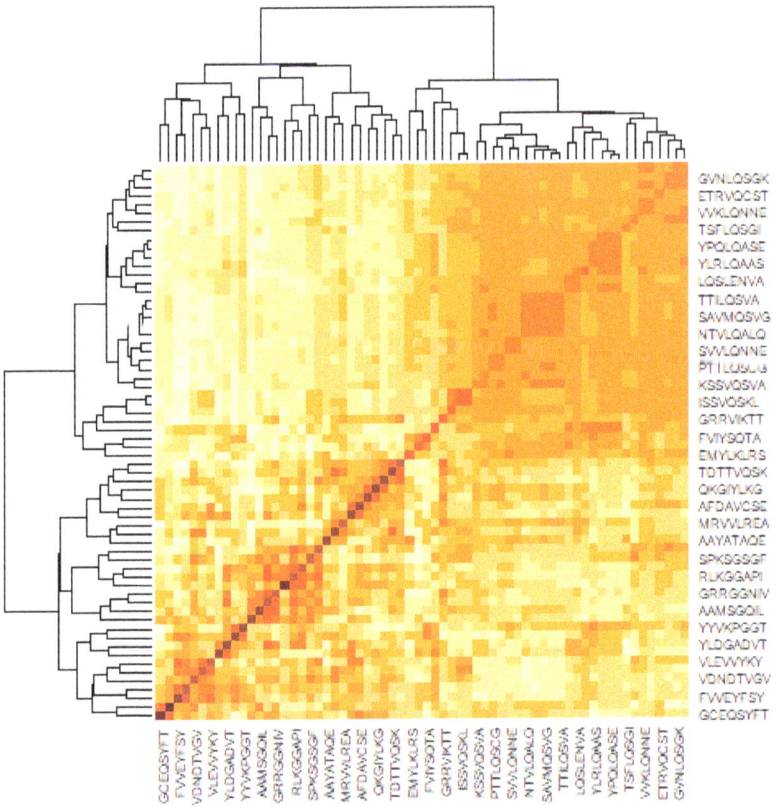

Figure 9.3. The correlation pattern between the hypothetical kernel peptides for the SARS peptide data.

into the top-right block demonstrate a very high level of mutual correlation. The outcome of this mutual correlation is that if a subset of the kernel peptides clustered into the top-right block is not selected for modeling, certain quantity of information may be lost leading to the low generalization capability of the bio-kernel machine model. If kPCA is employed, the correlation pattern between the PCs changes. Figure 9.4 shows this pattern. It can be seen that there is almost no correlation between PCs. This shows that ignoring some PCs, which are at the tail of the PC list to construct a bio-kernel machine model, will not lead to much information loss in terms of the data topological structure and hence the generalization capability.

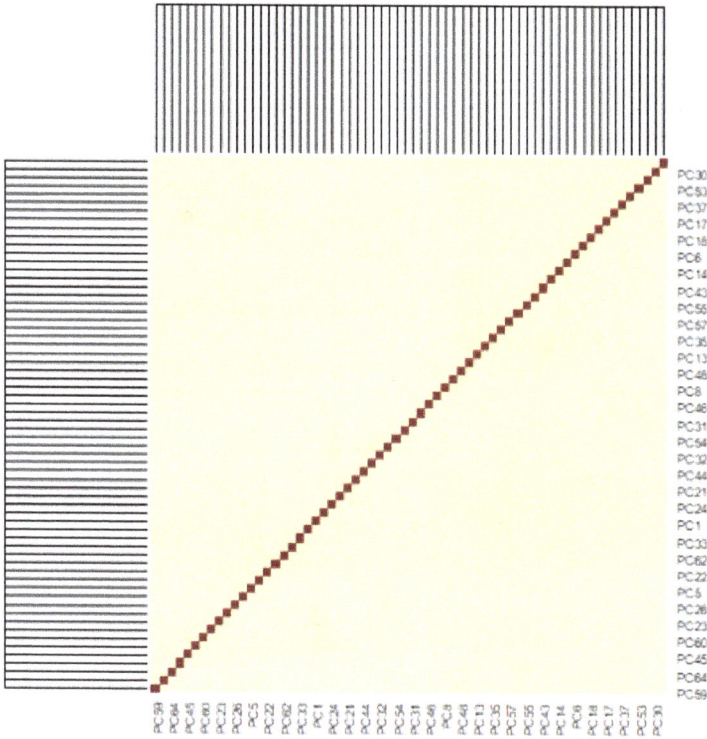

Figure 9.4. The correlation pattern between PCs for the SARS peptide data.

Besides the qualitative analysis shown here, we now examine the property quantitatively. Suppose the ratio of the variable variance against the whole correlation among the variables is calculated using the following equation:

$$\rho = \frac{\sum_{i=1}^{H} \sigma_i^2}{\sum_{i=1}^{H} \sum_{j \neq i}^{H} \sigma_{ij}} \tag{9.18}$$

where σ_{ij} is the correlation between two variables (the hypothetical kernel peptides or the principal components), σ_i^2 is the variance of a single variable, and H is the total number of variables. Table 9.1 shows the calculated ρ ratio for six peptide datasets. These quantitative data show that the variance of the hypothetical kernel peptides occupies a very small percentage of all data variance/

Bio-Kernel Machine and Its Applications

Table 9.1. The ρ ratios calculated for six peptide datasets. H stands for the ratio calculated in the original hypothesis space and P stands for the ratio calculated in the kPCA space.

Data name	H (%)	P (%)	References
O-linkage	5.6	100	Yang and Chou (2004b)
Factor Xa	3.2	97.9	Yang *et al.* (2003, 2006)
Caspase	8.1	100	Yang (2005b)
SARS	7.6	100	Yang (2005a)
HCV	3.8	100	Narayanan *et al.* (2002), Yang (2006)
HIV	11.0	100	Cai and Chou (1998), Yang *et al.* (2004), Yang and Thomson (2005)

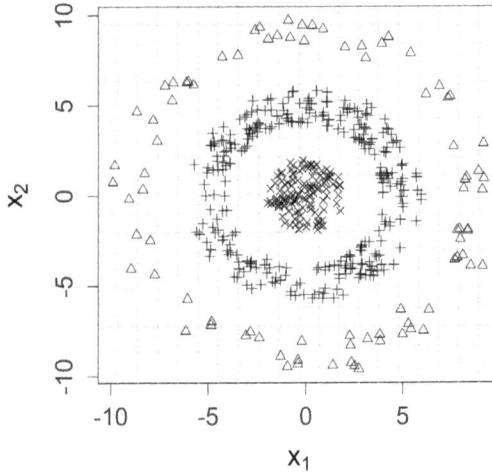

Figure 9.5. A nonlinear dataset with three classes of data points to show how kPCA works.

covariance, which is usually treated as the information of a dataset for pattern recognition through a learning process. Ignoring it may cause information loss in a model. However, the ρ ratios are very high, approaching 100% within the principal component space.

In addition to the orthogonality of a PCA model, what we are interested is the property of a kPCA model when the kernel trick is used. We now show how a kPCA model can map a nonlinear space to a linear space. Figure 9.5 shows a nonlinear dataset in a two-dimensional space, in which there are three classes of data points. Two conventionally employed kernel functions are applied to convert this nonlinearly separable space to a new space in which a kPCA model is constructed. The first one is called the polynomial kernel function which is shown in the following:

$$\mathcal{K}(\mathbf{x}_i, \mathbf{x}_j) = (\langle \mathbf{x}_i, \mathbf{x}_j \rangle + 1)^2 \qquad (9.19)$$

The second one is the so-called the radial basis kernel function which is shown in the following:

$$\mathcal{K}(\mathbf{x}_i, \mathbf{x}_j) = \exp\left(-\frac{(\mathbf{x}_i - \mathbf{x}_j)^2}{2\sigma^2}\right) \qquad (9.20)$$

Figure 9.6(a) shows the kPCA result using the polynomial kernel function and Figure 9.6(b) shows the kPCA outcome using the radial basis kernel function. It can be seen that both kPCA maps demonstrate that the new spaces (kPCA spaces) become linearly separable. However, it must be noted that it is not always true that a kernel function can transform a nonlinearly separable data space to a linearly separable space. For instance, if an improper σ^2 value is used for the radial basis kernel function, the kernel space may

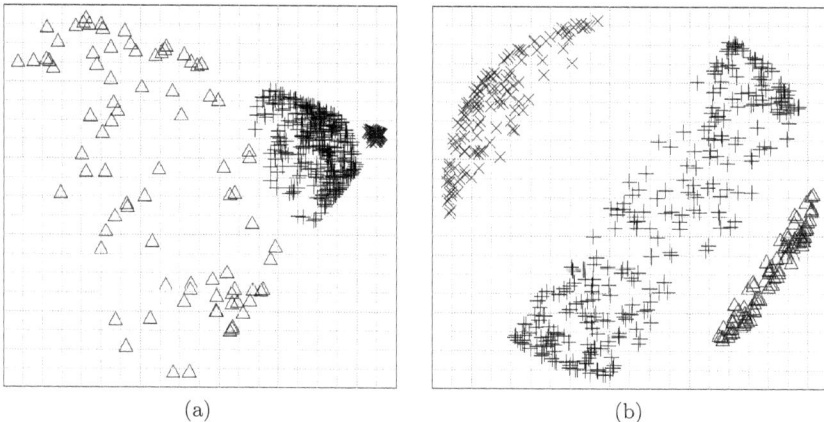

(a)   (b)

Figure 9.6. The kPCA models of two kernel spaces which employ two kernel functions. The original data are shown in Figure 9.5. (a) The use of the polynomial kernel function. (b) The use of the radial basis kernel function.

Bio-Kernel Machine and Its Applications

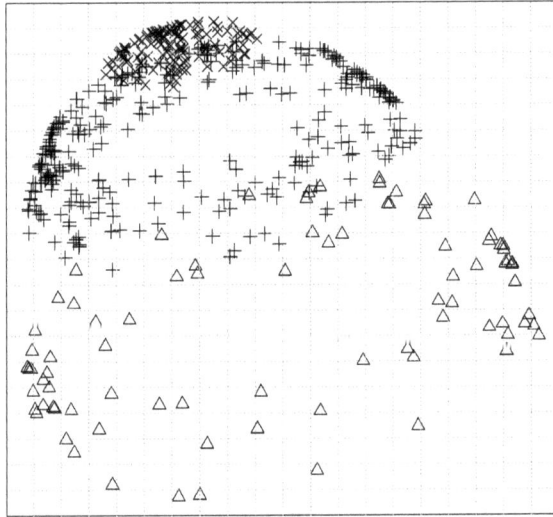

Figure 9.7. The kPCA model for the data space shown in Figure 9.5, where the radial basis kernel function was used and the σ^2 value was improperly set.

not be linearly separable or closer to linearly separable. Figure 9.7 shows a kernel space using the radial basis kernel function with an improperly selected σ^2 value. This kernel space is still heavily nonlinearly separable.

We now investigate two sets of bio-kPCA models to examine the linearity in the bio-kPCA space. The first set is constructed in the bio-kernel space formulated using a mutation matrix directly. This kernel space is referred to as the primary bio-kernel space. The second set is constructed using the polynomial kernel function shown in the following in the primary kernel space and is referred to as the secondary bio-kernel space:

$$\mathscr{K}(\mathbf{K}, \mathbf{K}) = (\langle \mathbf{K}, \mathbf{K} \rangle + 1)^2 \qquad (9.21)$$

Figure 9.8 shows the results of the abovementioned two types of bio-kPCA models. The Dayhoff mutation matrix was used for similarity measurement between HCV peptides, resulting in a kernel space for constructing a kPCA model. The visualization of the models was based on the first two PCs in either model. Through the comparison, it can be seen that the two models demonstrate a good

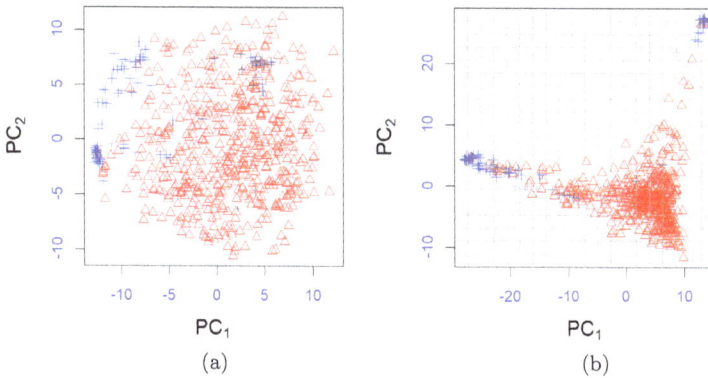

Figure 9.8. The bio-kPCA models constructed for the HCV dataset. (a) The bio-kPCA model constructed in the primary kernel space. (b) The bio-kPCA model constructed in the secondary kernel space. The triangles stand for the non-cleaved peptides and the pluses stand for the cleaved peptides.

separability between cleaved and non-cleaved peptides, especially the bio-kPCA model constructed in the secondary kernel space, which is shown in Figure 9.8(b). The models also demonstrate a very important property, i.e., the cleaved peptides are densely distributed into two corners, occupying a small area. This is as expected because the amino acids are normally randomly distributed in the non-cleaved peptides, while a small proportion of the amino acids are conserved in the cleaved peptides. The cleaved peptides are separated into two clusters. This may indicate that the cleaved peptides have two distinct amino acid composition patterns, which needs more investigation.

Figure 9.9 shows the eigenvalue distributions of the two models for the HCV dataset. It can be seen that the bio-kPCA model constructed in the secondary kernel space witnessed a quicker drop of the eigenvalues from the first few PCs.

Following this discussion, we now move on to the topic of the discrimination capability of a kPCA model. PCs have been used for discriminant analysis modeling for more than a decade (Jombart *et al.*, 2010; Yousefzadeh *et al.*, 2021; Bouhali *et al.*, 2023; Moraveikova *et al.*, 2023; Scaranto *et al.*, 2023). The advantage of discriminant analysis of principal components (DAPC) is that the

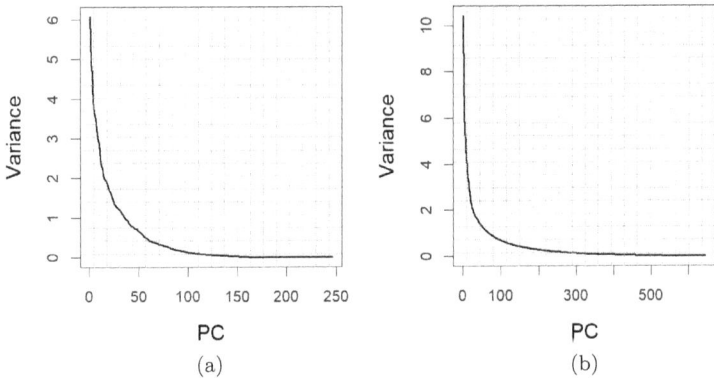

Figure 9.9. The eigen (variance) distribution of the bio-kPCA models constructed for the HCV dataset. (a) The bio-kPCA model constructed in the primary kernel space. (b) The bio-kPCA model constructed in the secondary kernel space.

variables (PCs) are mutually orthogonal, and hence such a model may only employ a few variables (PCs) to achieve a satisfactory and comparable generalization performance.

We now employ the DAPC approach to construct bio-kernel machine models for six peptide datasets. In general, there is no unique optimal number of PCs for all models, including the LS bio-kernel models and the deep bio-kernel machine (SVM bio-kernel) models. Some models may need a few PCs to approach the peak of the best performance, but other models may need to employ more PCs. Figure 9.10(a) shows this DAPC scenario for the Factor Xa dataset. It can be seen that the best performance for both LS bio-kernel models and the deep bio-kernel models are approached when the number of PCs is less than 30. This number is less than the total number of PCs. However, Figure 9.10(b) shows that the best number of PCs employed in the bio-kernel models for the HIV dataset is large. One needs to employ almost half of the PCs to approach the highest AUC values.

Here, we further analyze the correlation pattern between the bio-kernels. The pattern is based on the following ratio ϑ:

$$\vartheta = \sum_{h=1}^{H} \alpha_h \tag{9.22}$$

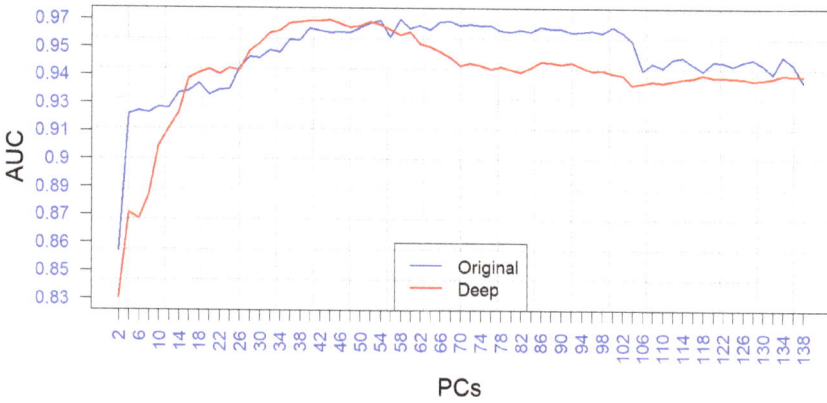

(a)

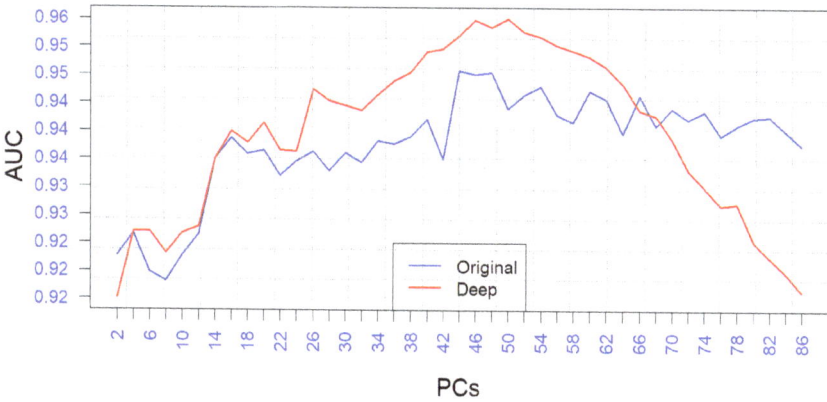

(b)

Figure 9.10. The DAPC of bio-kPCA models employing different numbers of PCs constructed for two datasets. (a) The Factor Xa dataset. (b) The HIV dataset. "Original" and "deep" stand for the LS and the deep (SVM) bio-kernel models, respectively.

where α_h is defined as follows:

$$\alpha_h = \frac{\sum_i \rho(\boldsymbol{v}_i, \boldsymbol{v}_h)}{\sigma_h^2} \tag{9.23}$$

In Eq. (9.23), σ_h^2 is the variance of the hth kernel and $\rho(\boldsymbol{v}_i, \boldsymbol{v}_h)$ is the covariance between hypothetical kernels $\boldsymbol{v}_i$ and $\boldsymbol{v}_h$. A great or small ϑ value indicates a high-level or low-level correlation between kernels. Figure 9.11 shows the distributions of the ϑ values across six

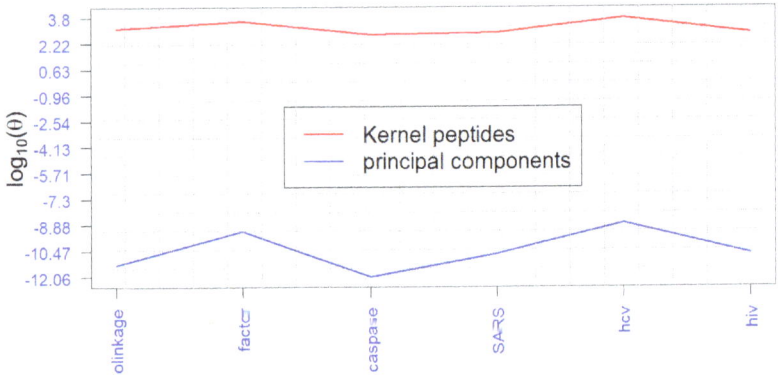

Figure 9.11.   The distributions of the ϑ values for kernels and PCs in six peptide datasets.

peptide datasets for both kernel peptides and PCs. Note that when calculating the ϑ values for the PCs for each dataset, the hypothetical kernels $\boldsymbol{v}_h$ are replaced by PCs. It can be seen that the original kernel peptides in these six peptide datasets are highly correlated, while the PCs derived from each dataset show almost no correlation due to orthogonalization. This is why a DAPC bio-kernel machine model does not need to employ all PCs for modeling a dataset if a subset of PCs have maintained a majority of the data variance.

We now carry out an experiment, in which two sets of simulations were implemented. In the first set of simulations, two to ten hypothetical kernel peptides were randomly selected. A bio-kernel machine was constructed followed by a measurement of AUC. This was repeated ten times to handle the uncertainty. Mean AUC and standard deviation AUC were recorded across ten models. In the meantime, a kPCA model was constructed in the kernel space, which is referred to as the second set of simulations. Afterward, two to ten PCs were selected to construct bio-kernel machines. AUC was measured at the same time. Table 9.2 shows the results. We name the first set of models as the reduced bio-kernel models, which do not use the PCs, and the second set as the reduced bio-kPCA or DAPC bio-kernel models. It can be seen that the DAPC bio-kernel models outperformed the reduced bio-kernel models significantly.

Table 9.2. The AUC values of the models constructed based on a reduced number of raw hypothetical kernel peptides and kPCA kernel peptides. Both the number of raw hypothetical kernel peptides and the number of kPCA kernel peptides were varied from two to ten. "Reduce" stands for the reduced bio-kernel models without using the PCs.

No		2	3	4	5	6	7	8	9	10
O-linkage	reduce	0.75	0.68	0.8	0.81	0.84	0.88	0.89	0.87	0.90
	DAPC	0.91	0.91	0.92	0.92	0.92	0.92	0.92	0.92	0.92
Factor Xa	reduce	0.64	0.75	0.79	0.84	0.86	0.87	0.89	0.90	0.88
	DAPC	0.87	0.91	0.91	0.91	0.91	0.91	0.91	0.91	0.91
Caspase	reduce	0.82	0.87	0.85	0.88	0.94	0.93	0.91	0.92	0.94
	DAPC	0.93	0.93	0.92	0.93	0.93	0.93	0.93	0.93	0.93
SARS	reduce	0.70	0.85	0.86	0.89	0.85	0.90	0.89	0.93	0.88
	DAPC	0.86	0.87	0.89	0.89	0.90	0.88	0.91	0.89	0.92
HCV	reduce	0.68	0.73	0.77	0.76	0.83	0.85	0.91	0.89	0.87
	DAPC	0.95	0.95	0.95	0.96	0.96	0.96	0.96	0.95	0.96
HIV	reduce	0.62	0.72	0.81	0.82	0.79	0.83	0.86	0.87	0.89
	DAPC	0.87	0.9	0.92	0.91	0.93	0.92	0.92	0.92	0.93

This is mainly because the PCs absorbed more information about the correlation between raw variables (kernel peptides).

Figure 9.12 shows the AUC measurements of the reduced bio-kernel models and the DAPC bio-kernel models constructed for the HCV dataset (Narayanan *et al.*, 2002; Yang, 2006). The models employed varying number of hypothetical kernel peptides and PCs. It can be seen that the DAPC bio-kernel models outperformed the reduced bio-kernel models, which used a randomly selected subset of hypothetical kernel peptides the whole way for this dataset. In particular, when employing two raw hypothetical kernel peptides, the mean AUC was about 0.68 for that dataset. However, the AUC of the DAPC bio-kernel models employing two PCs reached 0.95. This increase is no doubt very significant.

Table 9.3 shows the performance (AUC and MCC) of the DAPC bio-kernel models constructed for six peptide datasets. Figure 9.13 shows their ROC curves. First, the performance is better than most previous bio-kernel models, including the LS bio-kernel models, and is comparable to the deep bio-kernel machine models. This shows that the property of the bio-kernels is very important in

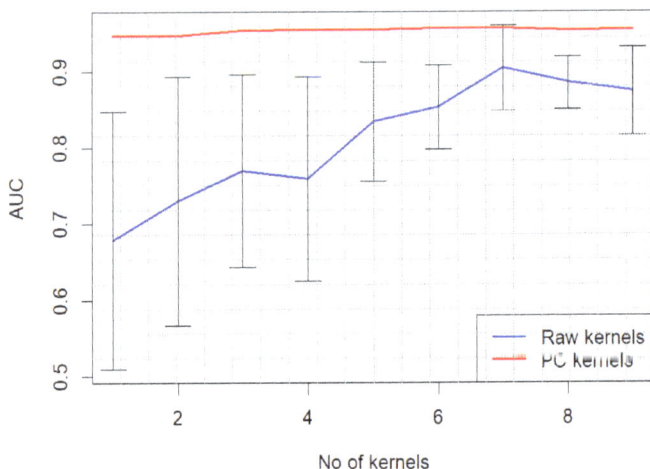

Figure 9.12. The AUC measured for the reduced kernel models constructed based on the varying number hypothetical kernel peptides and PCs for the HCV dataset. The line with the error bars stands for the performance measured on the reduced bio-kernel models built based on a randomly selected subset of hypothetical kernel peptides. The thick line on the top which has no error bar stands for the performance measured on the DAPC bio-kernel models.

Table 9.3. The AUC and MCC values of the DAPC models (for both the LS bio-kernel and the deep bio-kernel machine).

Data name	LS bio-kernel		SVM bio-kernel	
	AUC	MCC	AUC	MCC
O-linkage	0.91	0.63	0.92	0.64
Factor Xa	0.95	0.61	0.94	0.66
Caspase	0.96	0.73	0.97	0.84
SARS	0.94	0.71	0.98	0.87
HCV	0.98	0.65	0.97	0.7
HIV	0.95	0.66	0.96	0.7

constructing a better discriminative model and the orthogonality between bio-kernels can help generate parsimonious models with similar model performance. Second, some models require more PCs to be employed for a satisfactory performance, but some others may

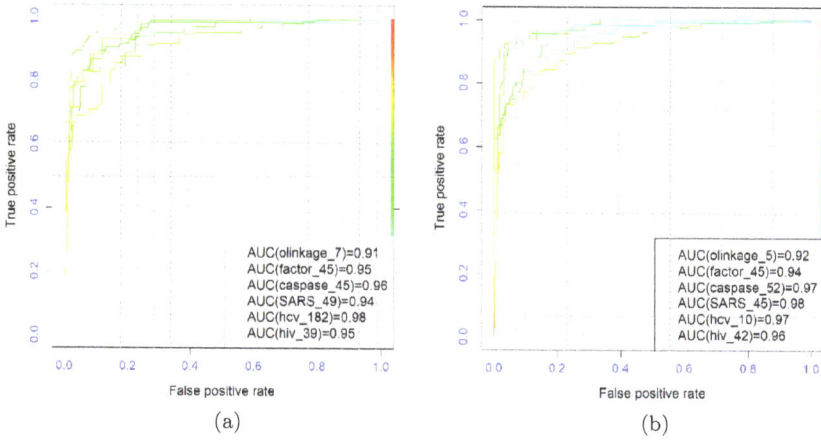

AUC(olinkage_7)=0.91
AUC(factor_45)=0.95
AUC(caspase_45)=0.96
AUC(SARS_49)=0.94
AUC(hcv_182)=0.98
AUC(hiv_39)=0.95

AUC(olinkage_5)=0.92
AUC(factor_45)=0.94
AUC(caspase_52)=0.97
AUC(SARS_45)=0.98
AUC(hcv_10)=0.97
AUC(hiv_42)=0.96

(a)   (b)

Figure 9.13. The ROC curves of the DAPC bio-kernel models constructed for six peptide datasets. The optimal number of PCs was employed in each model. The number on the right side of the dataset names is the optimal number of PCs. For instance, "olinkage_7" means that seven PCs were employed in that model. (a) The LS bio-kernel models. (b) The deep (SVM) bio-kernel models.

need very few. For instance, the deep bio-kernel machine model constructed for the O-linkage dataset only requires five PCs.

9.2 Kernelized self-organizing map

The kernel approach has also been used for enhancing the SOM algorithm for both unsupervised and supervised machine learning projects over two decades (MacDonald and Fyfe, 2000; Papadimitriou and Likothanassis, 2004; Laua *et al.*, 2006; Villa and Rossi, 2007; Chen *et al.*, 2009; Yu *et al.*, 2023), including the bio-kernel SOM (Yang and Young, 2005). The error function employed by SOM is shown in the following:

$$\varepsilon = \frac{1}{2}\|\mathbf{x}_n - \mathbf{w}_h\|^2 = \frac{1}{2}(\mathbf{x}_n^t\mathbf{x}_n + \mathbf{w}_h^t\mathbf{w}_h - 2\mathbf{x}_n^t\mathbf{w}_h) \qquad (9.24)$$

Suppose the vectors have been mean–variance normalized. We then have the following property:

$$\mathbf{x}_n^t\mathbf{x}_n = \langle \mathbf{x}_n, \mathbf{x}_n \rangle = 1 \qquad (9.25)$$

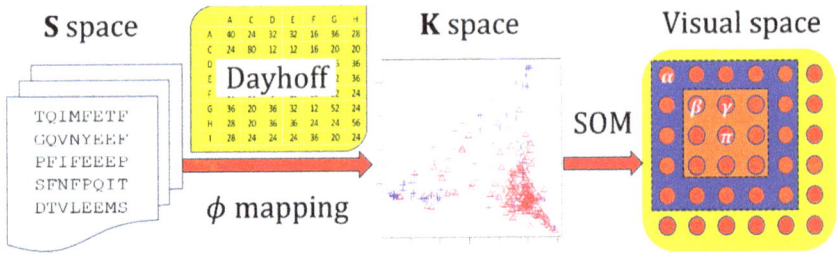

Figure 9.14   An illustration of a bio-kSOM model structure. The **S** space is composed of peptides. The **K** space is a kernel space. The mapping from the **S** space to the **K** space is completed by employing a mapping function based on a mutation matrix. Based on the **K** space, a visual space is generated using bio-kSOM.

Replacing $\mathbf{w}_h$ by $\mathbf{x}_h$ leads to a revision of the abovementioned error minimization equation to the following format:

$$\varepsilon = (1 - \langle \mathbf{x}_n, \mathbf{x}_h \rangle) \tag{9.26}$$

Because both $\mathbf{x}_n$ and $\mathbf{x}_h$ are mean–variance normalized, we then have the following relationship:

$$\langle \mathbf{x}_n, \mathbf{x}_h \rangle < 1 \tag{9.27}$$

This shows that replacing $\phi_{nh} = \langle \mathbf{x}_n, \mathbf{x}_h \rangle$ by a proper kernel function can lead to the construction of a kSOM model. If $\mathbf{x}_n$ is replaced by a training peptide $\mathbf{s}_n$ and $\mathbf{x}_h$ is replaced by a hypothetical kernel peptide $\mathbf{s}_h$, a bio-kSOM is thus structured. Figure 9.14 shows such a bio-kSOM model.

Figure 9.15 shows how bio-kSOM can be employed to visualize the topological structure of a bio-kernel model, which only employs the two most discriminative kernel peptides. One was constructed using the Dayhoff mutation matrix and the other was constructed using the BLOSUM62 mutation matrix. The selection of the two most discriminative kernel peptides for generating a topological map for a dataset using bio-kSOM was done using the t test. Both maps show that the two classes of peptides are well separated even only using the two kernel peptides which have the best discrimination capability, i.e., the lowest p values of the t test. It is as expected that the cleaved peptides are densely distributed in this map compared to the

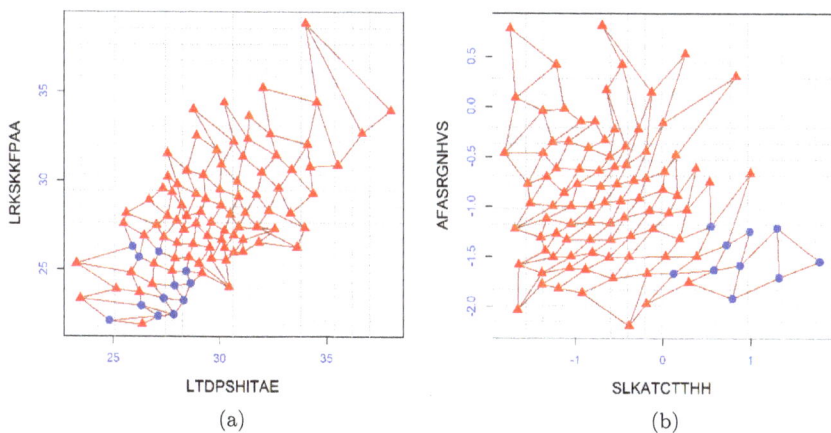

Figure 9.15. The visualization of the bio-kSOM models constructed for the HCV dataset using the two most discriminant kernel peptides. The filled triangles stand for the non-cleaved HCV peptides. The filled dots represent the cleaved HCV peptides. (a) The bio-kSOM model constructed using the Dayhoff mutation matrix. (b) The bio-kSOM model constructed using the BLOSUM62 mutation matrix.

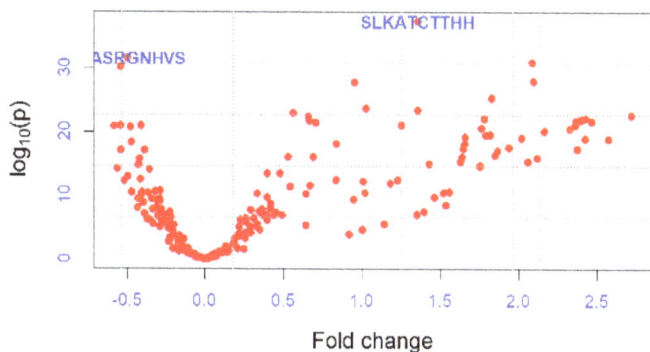

Figure 9.16. The distribution of the t test p values of the peptides with respect to the two most discriminant kernel peptides for the HCV dataset using the BLOSUM62 mutation matrix to align the peptides.

non-cleaved peptides. Figure 9.16 shows the distribution of the t test p values against the fold change for the bio-kSOM model constructed for the HCV dataset. The latter is the difference between the means of two types peptides based on the alignment scores calculated using the BLOSUM62 mutation matrix.

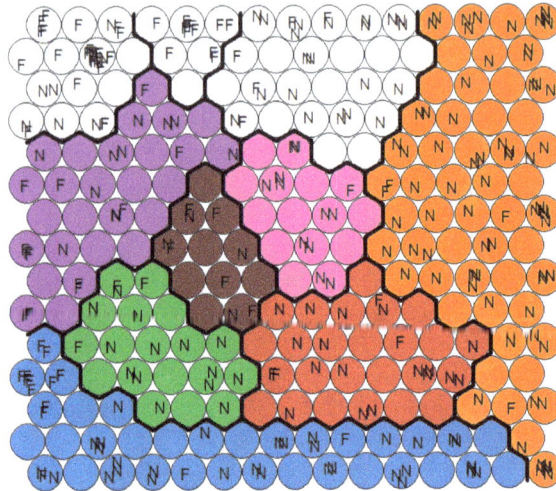

Figure 9.17. The bio-kSOM model of the HIV data with the number of output neurons, which was 80% of the number of HIV peptides. N stands for the non-cleaved peptides and F stands for the cleaved peptides. The neurons in different colors represent different clusters, which were made using a clustering algorithm.

Figure 9.17 shows a map of the bio-kSOM model constructed for the HIV dataset. In the model, the percentage of the output neurons over the number of peptides was 80%.

The original objective of SOM is for data visualization or topological structure reservation through unsupervised learning as discussed in Chapter 3. One important understanding in many applications is that the phenotypic pattern of a set of objects with the same background is closely correlated with the genotypic pattern of these objects. This means that the phenotypic pattern can be a function of the genotypic pattern. Thus, a discovered data structure by SOM may help the associative study between the genotypic pattern (the amino acid composition within the peptides) and the phenotypic pattern (the functional status such as cleavage or not of the peptides). A key question then arises: How can a model constructed by SOM be utilized for supervised learning such as discriminant analysis?

It is known that there is a difference between an SOM model and a PCA model. A PCA model delivers a quantitative model which has the same dimension as input data. Therefore, the PCs of a PCA model have a one-to-one correspondence with the input variables. We can, therefore, utilize the PCs, and hence DAPC, for discriminant analysis in a straightforward manner. However, the model parameter set **W** of an SOM model has no one-to-one corresponding relationship with the input variables. Therefore, we cannot use an approach similar to DAPC to build a discriminative model directly based on the weights of the SOM model.

Nevertheless, we still have two opportunities. We can utilize the SOM map to carry out discriminant analysis. For discriminant analysis, a confusion matrix is commonly employed for the evaluation. Suppose a confusion matrix as shown in the following:

$$C = \begin{pmatrix} \text{TN} & \text{FP} \\ \text{FN} & \text{TP} \end{pmatrix} \tag{9.28}$$

To generate such a confusion matrix, the following rules are used. First, if an output neuron is pure for non-functional peptides, all the peptides mapped to this neuron is counted to the measurement called TN. If an output neuron is pure for functional peptides, all the peptides mapped to this neuron is counted to the measurement called TP. The second situation occurs when the class labels of the peptides mapped to an output neuron are mixed, i.e., the output neuron is impure. Such an output neuron is referred to as a confusion neuron. If the number of functional peptides mapped to a confusion neuron is greater than the number of non-functional peptides mapped to the same confusion neuron, the number of functional peptides is counted to the measurement called TP and the number of non-functional peptides is counted to the measurement called FP. Otherwise, the number of non-functional peptides is counted to the measurement called TN and the number of functional peptides is counted to the measurement called FN. Using this approach, a confusion matrix can thus be generated from an SOM map after training.

But, there is another question. A free parameter of an SOM model is the number of output neurons. There is no doubt that different numbers of output neurons will demonstrate different performances. Too few output neurons will certainly enable data points of different classes to be mapped to the same output neuron, causing many

Table 9.4. The MCC measurements for the bio-kSOM models constructed for six peptide datasets with different percentages of output neurons. The percentages were set from 10% to 80%. Figures in bold stand for the best performance for each dataset.

	10%	20%	30%	40%	50%	60%	70%	80%
O-linkage	0.53	0.55	0.61	0.61	0.74	0.72	**0.80**	0.79
Factor Xa	0.66	0.72	0.74	0.74	0.82	**0.83**	0.80	0.83
Caspase	0.73	0.76	0.83	0.8	0.78	**0.92**	0.80	0.88
SARS	0.82	0.85	0.8	0.92	0.92	**0.95**	0.90	0.95
HCV	0.83	0.87	0.89	0.91	0.80	0.92	0.90	**0.93**
HIV	0.68	0.69	0.75	0.82	0.79	0.83	0.80	**0.86**

Table 9.5. The numbers of pure output neurons in the bio-kSOM models constructed for six peptide datasets with different numbers of output neurons. The percentages were set from 10% to 80%. Figures in bold stand for the greatest percentages of the pure output neurons.

	10%	20%	30%	40%	50%	60%	70%	80%
O-linkage	28	37	56	58	67	61	**73**	62
Factor Xa	56	62	66	66	**73**	64	64	57
Caspase	38	72	72	69	69	**75**	60	67
SARS	56	68	72	**84**	77	72	73	68
HCV	80	**85**	84	80	78	72	71	62
HIV	32	55	67	**74**	69	67	65	60

confusion neurons. However, too many output neurons will cause the occupied output neurons to be sparsely distributed in an SOM map. Due to this reason, different numbers of the output neurons have been employed to construct multiple SOM models for the purpose of discriminant analysis using a confusion matrix for each of six peptide datasets. Table 9.4 shows the MCC measurements calculated based on the confusion matrices of the SOM models with different ratios of output neurons (and hence different sizes for the SOM models) constructed for six peptide datasets. The ratio was defined as the percentage of the output neurons with regard to the total number of peptides. The MCC was calculated by generating a confusion matrix for each SOM model as mentioned earlier.

Table 9.6. The numbers of empty output neurons in the bio-kSOM models constructed for six peptide datasets with different numbers of output neurons. The percentages were set from 10% to 80%. Figures in bold stand for the greatest percentages of the empty output neurons.

	10%	20%	30%	40%	50%	60%	70%	80%
O-linkage	0	0	0	1	14	22	14	**25**
Factor Xa	0	5	11	17	15	27	27	**36**
Caspase	0	0	6	10	14	19	**27**	27
SARS	0	4	0	6	16	25	21	**29**
HCV	0	3	8	14	17	25	26	**36**
HIV	0	0	6	9	15	22	25	**34**

Table 9.5 shows the number of pure output neurons of a bio-kSOM model. It is not surprising that a model with a fewer number of output neurons tends to have a fewer number of pure output neurons and therefore more confusion neurons.

Table 9.6 shows the percentages of the output neurons which have no peptides mapped in the six bio-kSOM models. Obviously, when the number of output neurons increases, the number of empty output neurons increases as well. The increase of the empty neuron in an SOM model is not welcome due to computing memory and computing speed.

The abovementioned discussion only explores whether a model constructed by SOM has some discrimination capability in addition to its function of visualizing the internal data structure. The next discussion in this chapter is on how to construct a supervised bio-kSOM model for a peptide dataset. To do this, we thus employ the cross-validation approach to implement a supervised learning process to assess truly how good a bio-kSOM model is for a peptide dataset.

Table 9.7 shows the performance (AUC and MCC) of the bio-kSOM models constructed for six peptide datasets. Figure 9.18 shows the ROC curves of the bio-kSOM models constructed using the Dayhoff mutation matrix and the BLOSUM62 mutation matrix separately. Note that each model performance was a mean of 20 bio-kSOM models. This is because each bio-kSOM model starts from the randomization of model parameters (weights). The model

Table 9.7. The AUC and MCC values for the bio-kSOM models constructed for six peptide datasets. Dayhoff stands for the models constructed using the Dayhoff mutation matrix and BLOSUM62 stands for the models constructed using the BLOSUM62 mutation matrix.

	Dayhoff		BLOSUM62	
Data name	AUC	MCC	AUC	MCC
O-linkage	0.90	0.63	0.88	0.58
Factor Xa	0.94	0.68	0.94	0.68
Caspase	0.96	0.78	0.90	0.88
SARS	0.97	0.86	0.97	0.89
HCV	0.97	0.81	0.96	0.82
HIV	0.89	0.65	0.91	0.67

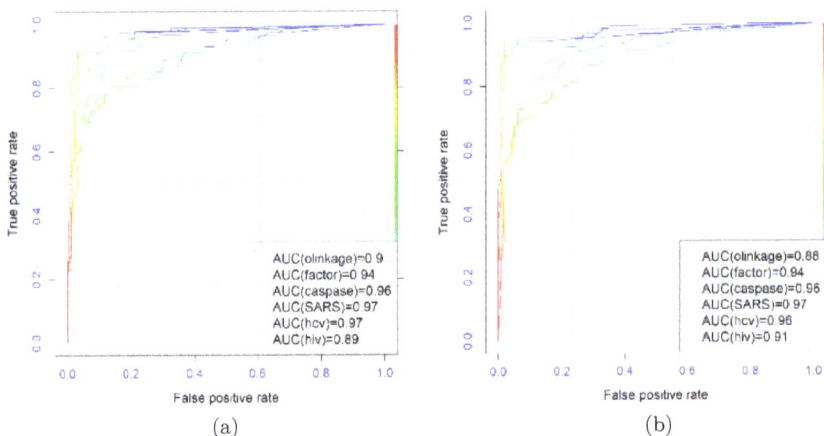

Figure 9.18. The ROC curves of the bio-kSOM models constructed for six peptide datasets. (a) The bio-kSOM models which employ the Dayhoff mutation matrix. (b) The bio-kSOM models which employ the BLOSUM62 mutation matrix.

performance is not compatible with most of the previous models. This is because SOM does not utilize the known class labels for training a model. Instead, it purely relies on the exploration of the internal data structure for discrimination of phenotypic variables.

The final discussion in this chapter is about how to develop a model which is similar to deep learning. The principle of such

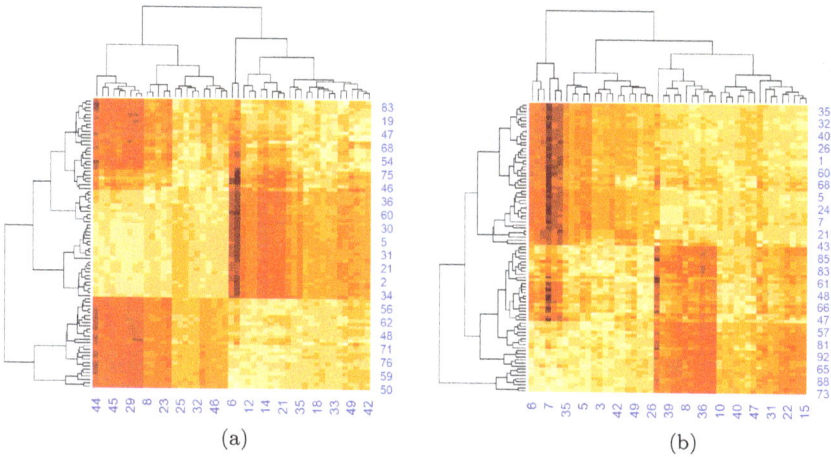

Figure 9.19. The heatmaps of the weight spaces of the bio-kSOM models. (a) The SARS model. (b) The Caspase model.

a deep model is to utilize the information of the weight space of a well-trained bio-kSOM model. There is no doubt that a well-trained SOM model should be able to reserve the topological structural information of a dataset by which the SOM model is constructed. Figure 9.19 shows two heatmaps of the weight spaces of two bio-kSOM models constructed for two peptide datasets (the SARS dataset and Caspase dataset). It can be seen that the weight spaces are well structured. The two heatmaps have delivered a very strong message, i.e., both the spaces are well clustered into blocks. This kind of non-random structure demonstrates that the weight spaces contain strong discriminative information. This is consistent with the learning mechanism of SOM because a mapping process using SOM is based on the following equation, which has been discussed in Chapter 3:

$$\pi_n = \operatorname*{argmin}_{m \in [1,M]} \left\{ \frac{1}{2} \|\mathbf{x}_n - \mathbf{w}_h\|^2 \right\} \tag{9.29}$$

This equation indicates that all the input data will be mapped to different locations of an SOM model according to the internal data structure. Peptides with a similar amino acid composition pattern will be mapped to the same or nearby output neurons. This mapping is guided by the SOM parameters. In other words, the data structure information has been well distributed into the weight space. To be more specific, some subspaces of the weight space will lead peptides to be mapped to the output neurons of non-cleaved peptides, while other subspaces of the weight space will lead peptides to be mapped to the output neurons of cleaved peptides. It has been analyzed in Chapter 6 that functional (cleaved or posttranslational modified) peptides have well reserved amino acid composition patterns, but non-functional peptides show a random composition pattern of amino acids. Therefore, a well-trained bio-kSOM model will have a weight space which possesses the discriminative power to map these two types of peptides into different areas of a bio-kSOM map.

Following this discussion, we can further examine how the weight space of a well-trained bio-kSOM model can be used for a supervised learning task. Figure 9.20 shows such a deep network of exploring the pattern stored in the weight space of a well-trained bio-kSOM model for the discrimination analysis of a peptide data. This deep network is referred to as discriminant analysis of the weight space of a SOM model (DASOM). In a DASOM model, the new space $\tilde{\mathbf{K}}_{som}$

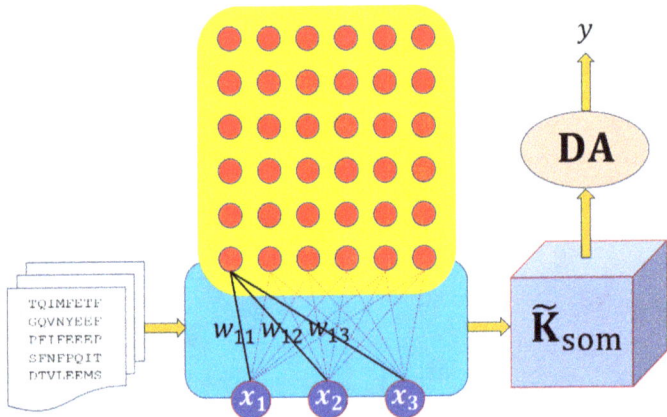

Figure 9.20. The diagram of DASOM. After a bio-kSOM model has been constructed, the training and testing peptides are mapped to a new space $\tilde{\mathbf{K}}_{som}$ through the weight space of a trained bio-kSOM model. A discriminant analysis process is carried out in this new space $\tilde{\mathbf{K}}_{som}$.

Table 9.8. The AUC and MCC values of the DASOM models constructed for six peptide datasets.

Data name	Regression		SVM	
	AUC	MCC	AUC	MCC
O-linkage	0.91	0.63	0.92	0.66
Factor Xa	0.93	0.61	0.93	0.74
Caspase	0.96	0.83	0.94	0.81
SARS	0.96	0.84	0.95	0.76
HCV	0.97	0.78	0.98	0.81
HIV	0.94	0.68	0.92	0.75

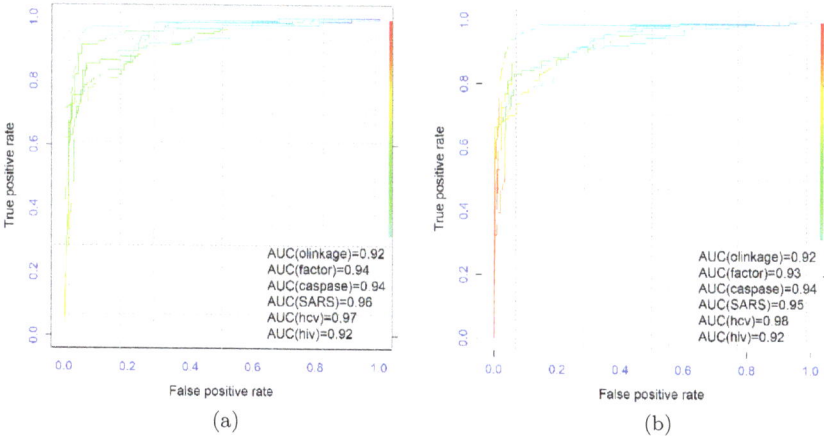

Figure 9.21. The ROC curves of the DASOM models constructed for six peptide datasets. (a) Regression models. (b) SVM models.

is formulated using the following format:

$$\tilde{k}_n = (\|k_n - \mathbf{w}_1\|^2, \|k_n - \mathbf{w}_2\|^2, \ldots, \|k_n - \mathbf{w}_M\|^2) \qquad (9.30)$$

The new space is structured as follows:

$$\tilde{\mathbf{K}}_{\text{som}} = (\tilde{k}_n)_{n=1}^N \qquad (9.31)$$

Based on $\tilde{\mathbf{K}}_{\text{som}}$, discriminant analysis is carried out using both the regression mode and the SVM mode. Table 9.8 shows the performance evaluation of two sets of the DASOM models constructed for six peptide datasets. Figure 9.21 shows the ROC curves of these models. It can be seen that the performance is very good, meaning that the weight space of a well-trained bio-kSOM model does contain rich information for discriminant analysis; this proves again that the internal data structure of a peptide dataset does have a close relationship with the phenotypic feature of the peptide dataset.

There are also progress in the research of the kernelized self-organizing map algorithm, which treats each output neuron as a Gaussian processor, hence there is a link between the kernel self-organizing map and Gaussian mixture models (Yin and Allinson, 2001; van Hulle, 2002; Yin, 2006).

9.3 Summary

This chapter has introduced some classical approaches to visualize a bio-kernel machine model. These two classical approaches are the principal component analysis and the self-organizing map algorithm. Both algorithms are unsupervised with different learning mechanisms and statistical backgrounds. The principal component analysis algorithm can orthogonalize an original data space to minimize the correlation between variables. In this way, the principal components, which are a linear function of the original variables (kernel peptides in this book), can fully represent the information of the original variables. In most situations, the first few principal components can retain a majority of the variance (hence information) stored in the original data space. Due to this important property, we can employ the first few principal components to construct a discriminant analysis model with minimized information loss or maintain a satisfactory discriminant power. Therefore, the research of discriminant analysis of principal components has great power in many real-world applications. In this chapter, we have shown that the discriminant analysis of bio-kernelized principal components can be very useful to build up a discriminant analysis model with a comparable and satisfactory generalization performance.

The self-organizing map algorithm searches for a codebook, which is a set of vectors, to represent the original vectors. The most important feature of the algorithm is its online competitive learning mechanism. The algorithm was initially developed mainly for topological reservation research. Because there is always some natural association between the internal (genotypical) data structure and the external (phenotypical) outcome for the objects with the same background, no matter whether this association is simple or complex, research into this type of association has been exercised in machine learning, and unsupervised learning regarding the internal data structure can therefore provide a foundational knowledge for supervised learning exercises. This is why the property of the self-organizing map has been well utilized for constructing some supervised learning models. Moreover, along with the most recent developments, kernelized self-organizing map algorithms have shown capability in constructing a model which has both visualization capability and supervised learning power. This chapter has therefore introduced two versions of supervised bio-kernelized self-organizing map algorithms and their applications to the peptide data study with success, with one version that utilizes the topological structure of the weight space of a well-trained bio-kernelized self-organizing map model.

Chapter 10

Future Developments

Bio-kernel machines have been developed and researched for many years. They have been enhanced from the original least-squares regression mode to some advanced modes, such as the orthogonal bio-kernel machine, the Bayesian bio-kernel machine, the intelligent bio-kernel machine, and the deep bio-kernel machine. Bio-kernel machines have been applied to different peptide data modeling and analysis projects, such as protease cleavage site prediction and post-translational modification site prediction. The basis is the mapping of a peptide space to a kernel space through the employment of a mutation matrix, by which a non-numerical peptide space is transferred to a numerical kernel space. The kernel space is supported by a set of kernel peptides, which are normally post-translational modified peptides or cleaved peptides. This is based on an essential property of these kinds of peptides; that is, most functional (cleaved or post-translational modified) peptides have a conserved amino acid composition pattern. The pattern originates from the protein structure's nature. Both protease cleavage and post-translational modification must ensure proper binding of two molecules to each other. Without solid structure conservation, binding of two molecules may not be possible. Therefore, using functional peptides to serve as kernels can ensure the establishment of a robust and reliable kernel space for data analysis, such as the discrimination analysis between functional and non-functional peptides.

Several questions still await better solutions. First, how can we be sure that a specific mutation matrix is the best? Among many different mutation matrices, selecting the best one for a specific dataset is a non-trivial task as discussed in this book. One reason is that some were developed earlier, such as the Dayhoff mutation matrix, while some were developed later. The other reason is that each mutation matrix was developed based on different scenarios, such as different sets of protein sequences as well different approaches, either statistical or algorithmic. If the performance variation is insignificant among models constructed using different mutation matrices, we may not bother about selecting the best one. However, as shown in this book, this is not the case. Therefore, some advanced techniques have been introduced in this book, such as the employment of the fusion technology and data-driven estimation of a mutation matrix from a specific dataset. However, how many mutation matrices should be included in a fusion process for analyzing a specific dataset? Certainly, some may be very non-informative or useless compared to others when analyzing different datasets. As may be obvious, a mean estimation may be significantly affected by an outlier in a dataset. Therefore, the following question arises: By which criterion should a mutation matrix be included or removed from a fusion process? No doubt, a more reliable and robust fusion technology should be researched to enhance the generalization capability of bio-kernel machine fusion in terms of the available mutation matrices. It is shown in this book that estimating a mutation matrix directly from a specific peptide dataset under investigation is possible. This can avoid the uncertainty when we have to select one mutation matrix to analyze a peptide dataset or make a decision about which mutation matrices can be included in a bio-kernel machine fusion process. But, a common observation is that the size of a peptide dataset may not be statistically significant. This means that new research is required to address the issue of how to efficiently revise a mutation matrix, which is estimated based on old peptides when new peptides are collected in the future. Therefore, all these issues need deeper research in the future.

Second, how can we be sure that a specific bio-kernel machine is the best? This book has introduced not only the original least-squares bio-kernel machine but also several advanced ones. Each has its own strengths, such as being a parsimonious model or being able to enhance the generalization capability. For instance, the orthogonal bio-kernel machine can generate a model which employs fewer variables (bio-kernels) to approach a satisfactory and comparable generalization performance. However, its generalization performance is not improved significantly, although such a model has a good interpretation power it has been constructed. On the other hand, by introducing more computational cost, deep bio-kernel machines can improve the generalization performance significantly. The short-coming of a deep bio-kernel machine model is its heavier model structure, which leads to lower efficiency of prediction, especially as the model is still partially treated as a black box in the machine learning community. Therefore, a more powerful learning machine may need to be researched so as to fulfill as many criteria as possible.

Besides the abovementioned issues, if combined with visualization algorithms, the interpretation capability of a bio-kernel machine is further enhanced when analyzing peptide data. This book has shown that kernelized principal component analysis and the kernelized self-organizing map have a natural space to embed the bio-kernel machine for both unsupervised and supervised peptide pattern discovery and analysis. This book has also introduced discriminant analysis of principal components and discriminant analysis of the self-organizing map, which work well, but the uncertainty regarding model sizes still remains and needs to be researched further.

The final point is whole-sequence data analysis. Limited by space, this issue has not been discussed in this book. But, the bio-kernel machine has the potential to construct a predictive model for protein function prediction through whole-sequence comparison by focusing on mutation positions or blocks only. Suppose a set of closely related protein sequences have been initially aligned. Focusing on the most variant blocks and constructing a bio-kernel model based on the mutations within these blocks may make the function prediction for a novel protein sequence more accurate and efficient. Moreover, as discussed in this book, the mismatch kernel can allow certain numbers of mismatches within a word. Rather than calculating the number of mismatches, constructing a bio-kernel machine model based on the employment of a mutation matrix may enhance the reliability and robustness of such a predictive model.

References

Aizerman, M. A., Braverman, E. M. and Rozonoer, L. I. (1964). Theoretical foundations of the potential function method in pattern recognition learning, *Automation and Remote Control*, 25, pp. 821–837.

Alam, S. I., Bansod, S. and Singh, L. (2008). Immunization against clostridium perfringens cells elicits protection against clostridium tetani in mouse model: identification of cross-reactive proteins using proteomic methodologies, *BMC Microbiology*, 8, pp. 194.

Altschul, S. F., Gish, W., Miller, W., Myers, E. W. and Lipman, D. J. (1990). Basic local alignment search tool, *Journal of Molecular Biology*, 215, pp. 403–410.

Atassi, M. Z. (1966). Significance of the amino acid composition of proteins. 1. Composition of hemoglobins and myoglobins in relation to their structure, function and evolution, *Journal of Theoretical Biology*, 11, pp. 227–241.

Azarya-Sprinzak, Z., Naor, D., Wolfson, H. J. and Nussinov, R. (1997). Interchanges of spatially neighbouring residues in structurally conserved environments, *Protein Engineering*, 10, pp. 1109–1122.

Baudat, G. and Anouar, F. (2000). Generalized discriminant analysis using a kernel approach, *Neural Computation*, 12, pp. 2385–2404.

Badalassi, V., Sircar, A., Solberg, J. M., Bae, J. W., Borowiec, K., Huang, P., Smolentsev, S. and Peterson, E. (2023). FERMI: fusion energy reactor models integrator, *Fusion Science and Technology*, 79, pp. 345–379.

Beck, Z. Q., Hervio, L., Dawson, P. E., Elder, J. H. and Madison, E. L. (2000). Identification of efficiently cleaved substrates for HIV-1 protease using a phage display library and use in inhibitor development, *Virology*, 274, pp. 391–401.

Ben-Hur, A. and Brutlag, D. L. (2003). Remote homology detection: a motif-based approach, *Bioinformatics*, 19, pp. 26–33.

Bengio, Y., LeCun, Y. and Hinton, G. (2015). Deep learning, *Nature*, 521, pp. 436–444.

Benner, S. A., Cohen, M. A. and Gonnet, G. H. (1994). Amino acid substitution during functionally constrained divergent evolution of protein sequences, *Protein Engineering*, 7, pp. 1323–1332.

Bishop, C. (1991). Improving the generalization properties of radial basis function neural networks, *Neural Computation*, 3, pp. 579–588.

Bishop, C. (1996). *Neural Networks for Pattern Recognition*, Oxford University Press.

Bishop, C. (2006). *Pattern Recognition and Machine Learning*, Springer-Verlag, New York.

Bjorck, A. (1967). Solving linear least-squares problems by Gram-Schmidt orthogonalization, *BIT Numerical Mathematics*, 7, pp. 1–21.

Blake, J. D. and Cohen, F. E. (2001). Pairwise sequence alignment below the twilight zone, *Journal of Molecular Biology*, 307, pp. 721–735.

Bors, A. G. and Pitas, I. (1996). Median radial basis function neural network, *IEEE Transactions on Neural Networks*, 7, pp. 1351–1364.

Bouhali, A., Homrani, A., Ferrand, N., Lopes, S. and Emam, A, M. (2023). Assessment of genetic diversity among native Algerian rabbit populations using microsatellite markers, *Archives Animal Breeding*, 66, pp. 207–215.

Briscoe, A. D. (2001). Functional diversification of lepidopteran opsins following gene duplication, *Molecular Biology and Evolution*, 18, pp. 2270–2279.

Branlant, C., Krol, A., Ebel, J. P., Lazar, E., Haendler, B. and Jacob, M. (1982). U2 RNA shares a structural domain with U1, U4, and U5 RNAs, *EMBO Journal*, 1, pp. 1259–1265.

Breiman, L. (2001). Random forests, *Machine Learning*, 45, pp. 5–32.

Breiman, L., Friedman, J. H., Olshen, R. A. and Stone, C. J. (1984). *Classification and regression trees*, Wadsworth & Brooks/Cole Advanced Books & Software, Monterey, CA.

Broomhead, D. S. and Lowe, D. (1988). Radial basis functions, multi-variable functional interpolation and adaptive networks, *Memorandum Report at Royal Signal and Radar Establishment Malvern*.

Buerkner, P. (2017). brms: An R package for bayesian multilevel models using Stan, *Journal of Statistical Software*, 80, pp. 1–28.

Bures, J. and Larrosa, I. (2023). Organic reaction mechanism classification using machine learning, *Nature*, 613, pp. 689–689.

Burton, D. K., Shore, J. E. and Buck, J. T. (1983). A generalization of isolated word recognition using vector quantization, *IEEE International Conference on Acoustics Speech and Signal Processing*, 8, pp. 1021–1024.

Butler, K. T., Davies, D. W., Cartwright, H., Isayev, O. and Walsh, A. (2018). Machine learning for molecular and materials science, *Nature*, 559, pp. 547–555.

Buzon, M. J., Marfil, S., Puertas, M. C., Garcia, E., Clotet, B., Ruiz, L., Blanco, J., Martinez-Picado, J. and Cabrera, C. (2008). Raltegravir susceptibility and fitness progression of HIV type-1 integrase in patients on long-term antiretroviral therapy, *Antiviral Therapy*, 13, pp. 881–893.

Cai, Y. D. and Chou, K. C. (1998). Artificial neural network model for predicting HIV protease cleavage sites in protein, *Advances in Engineering Software*, 29, pp. 119–128.

Campbell, D. T. (1960). Blind variation and selective retention in creative thoughts as in other knowledge processes, *Psychological Review*, 67, pp. 380–400.

Carroll, R. J. and Ruppert, D. (1996). The use and misuse of orthogonal regression in linear errors-in-variables models, *The American Statistician*, 50, pp. 1–6.

CBN (IUPAC-IUB Commission on Biochemical Nomenclature) (1968). A one-letter notation for amino acid sequences, *European Journal of Biochemistry*, 5, pp. 151–153.

Charlesworth, M. J. (1956). Aristotle's Razor, *Philosophical Studies*, 6, pp. 105–112.

Chiang, J. H. and Ho, S. H. (2008). A combination of rough-based feature selection and RBF neural network for classification using gene expression data, *IEEE Transactions on Nanobioscience*, 7, pp. 91–99.

Chen, J., Holliday, J. and Bradshaw, J. (2009). A machine learning approach to weighting schemes in the data fusion of similarity coefficients, *Journal of Chemical Information and Modeling*, 49, pp. 185–194.

Chen, S, Cowan, C. N. and Grant, P. M. (1991). Orthogonal least-squares learning algorithm for radial basis function networks, *IEEE Transactions on Neural Networks*, 2, pp. 302–309.

Chou, K. C. (2011). Some remarks on protein attribute prediction and pseudo amino acid composition. *Journal of Theoretical Biology*, 273, pp. 236–247.

Christabel, G. J. and Subhajini, A. C. (2023). KPCA-WRF-prediction of heart rate using deep feature fusion and machine learning classification with tuned weighted hyper-parameter, *Network*, 3, pp. 1–32.

Ciuonzo, D., Papa, G., Romano, G., Salvo Rossi, P. and Willett, P. (2013). One-bit decentralized detection with a rao test for multisensor fusion, *IEEE Signal Processing Letters*, 20, pp. 861–864.

Cofe, H. C. F., Brumme, Z. L. and Harrigan, P. R. (2001). Human immunodeficiency virus type 1 protease cleavage site mutations associated with protease inhibitor cross-resistance selected by indinavir, ritonavir, and/or saquinavir, *Journal of Virology*, 75, pp. 589–594.

Cortes, C. and Vapnik, V. N. (1995). Support-vector networks, *Machine Learning*, 20, pp. 273–297.

Cottier, V., Barberis, A. and Luthi, U. (2006). Novel yeast cell-based assay to screen for inhibitors of human cytomegalovirus protease in a high-throughput format, *Antimicrob Agents Chemother*, 50, pp. 565–567.

Crooks, G. E., Green, R. E. and Brenner, S. E. (2005). Sequence analysis pairwise alignment incorporating dipeptide covariation, *Bioinformatics*, 21, pp. 3704–3710.

Dayhoff, M. O. and Schwartz, R. M. (1978). A model of evolutionary change in proteins, *Atlas of Protein Sequence and Structure*, 5, pp. 345–352.

Dempster, A. P., Laird, N. M. and Rubin, D. B. (1977). Maximum likelihood from incomplete data via the EM Algorithm, *Journal of the Royal Statistical Society, Series B*, 39, pp. 1–3.

Dhillon, I. S., Guan, Y. and Kulis, B. (2004). Kernel k-means, spectral clustering and normalized cuts, *Proceedings of 10th ACM Knowledge Discover and Data Mining Conference*, pp. 551–556.

Dhillon, I. S., Guan, Y. and Kulis, B. (2007). Weighted graph cuts without eigenvectors: a multilevel approach, *IEEE Transactions on Pattern Analysis and Machine Intelligence*, 29, pp. 1944–1957.

Digianantonio, K. M., Korolev, M. and Hecht, M. H. (2017). A non-natural protein rescues cells deleted for a key enzyme in central metabolism, *ACS Synthetic Biology*, 6, pp. 694–700.

Dosztanyi, Z. and Torda, A. E. (2001). Amino acid similarity matrices based on force fields, *Bioinformatics*, 17, pp. 686–699.

Dowd, A. (2023). Elucidating cellular metabolism and protein difference data from DIGE proteomics experiments using enzyme assays, *Methods in Molecular Biology*, 2596, pp. 399–419.

Dowe, D. L., Gardner, S. and Oppy, G. (2007). Bayes not bust! Why simplicity is no problem for Bayesians, *British Journal for the Philosophy of Science*, 58, pp. 709–754.

Drapeau, G. R. (1978). The primary structure of taphylococcal protease, *Canadian Journal of Biochemistry and Physiology*, 56, pp. 534–544.

Duda, R. O., Hart, P. E. and Stork, D. G. (2000). *Pattern Analysis*, Wiley-Interscience, New York.

Dulos, J., Kaptein, A., Kavelaars, A., Heijnen, C. and Boots, A. (2005). Tumour necrosis factor-alpha stimulates hydro-epiandrosterone metabolism in human fibroblast-like synoviocytes: a role for nuclear factor-kappaB and activator protein-1 in the regulation of expression of cytochrome p450 enzyme 7b, *Arthritis Research & Therapy*, 7, pp. R1271–R1280.

Easteal, S. and Collet, C. (1994). Consistent variation in amino-acid substitution rate, despite uniformity of mutation rate: protein evolution in mammals is not neutral, *Molecular Biology and Evolution*, 11, pp. 643–647.

Eddy, S. R. (2004). Where did the BLOSUM62 alignment score matrix come from? *Nature Biotechnology*, 22, pp. 1035–1036.

Eduardo, P. C. and Danchin, R. A. (2004). An analysis of determinants of amino acids substitution rates in bacterial proteins, *Molecular Biology and Evolution*, 21, pp. 108–116.

Efron, B., Hastie, T., Johnstone, I. and Tibshirani, R. (2004). Least angle regression, *Annals of Statistics*, 32, pp. 407–499.

Elbe, S. and Buckland-Merrett, G. (2017). Data, disease and diplomacy: GISAID's innovative contribution to global health, *Global Challenges*, 1, pp. 33–46.

Erawijantari, P. P., Mizutani, S., Shiroma, H., Shiba, S., Nakajima, T., Sakamoto, T., Saito, Y., Fukuda, S., Yachida, S. and Yamada, T. (2020). Influence of gastrectomy for gastric cancer treatment on faecal microbiome and metabolome profiles, *Gut*, 69, pp. 1404–1415.

Estrada, E. (2020). Topological analysis of SARS CoV-2 main protease, *Chaos*, 30, pp. 061102.

Felsenstein, J. (1981). Evolutionary trees from DNA sequences: a maximum likelihood approach, *Journal of Molecular Evolution*, 17, pp. 368–376.

Fisher, R. A. (1936). The use of multiple measurements in taxonomic problems, *Annals of Eugenics*, 7, pp. 179–188.

Friedhelm, S., Kestler, H. A. and Palm, G. (2001). Three learning phases for radial-basis-function networks, *Neural Networks*, 14, pp. 439–458.

Garpebring, A., Brynolfsson, P., Kuess, P., Georg, D., Helbich, T. H., Nyholm, T. and Löfstedt, T. (2018). Density estimation of grey-level co-occurrence matrices for image texture analysis, *Physics in Medicine & Biology*, 63, pp. 195017.

Garey, M. R. and Johnson, D. S. (1979). *Computers and Intractability: A Guide to the Theory of NP-Completeness*, W.H. Freeman.

Gaston, H., Gonnet, M., Cohen, A. and Benner, S. (1992). Exhaustive matching of the entire protein sequence database, *Science*, 256, pp. 1443–1445.

Gatanaga, H., Suzuki, Y., Tsang, H., Tang, C. and summers, M. F. (2002). Amino acid substitutions in gag protein at non-cleavage sites are indispensable for the development of a high multitude of HIV-1 resistance against protease inhibitors, *Journal of Biological Chemistry*, 277, pp. p5952–p5961.

Gazzola, S., Novati, P. and Russo, M. R. (2015). On Krylov projection methods and Tikhonov regularization, *Electronic Transactions on Numerical Analysis*, 44, pp. 83–123.

Ghanty, P. and Pal, N. R. (2009). Prediction of protein folds: extraction of new features, dimensionality reduction, and fusion of heterogeneous classifiers, *IEEE Transactions on Nanobioscience*, 8, pp. 100–110.

Ghosh-Dastidar, S., Adeli, H. and Dadmehr, N. (2008). Principal component analysis-enhanced cosine radial basis function neural network for robust epilepsy and seizure detection, *IEEE Transactions on Biomedical Engineering*, 55, pp. 512–518.

Gilmour, S. G. and Goos, P. (2009). Analysis of data from non-orthogonal multistratum designs in industrial experiments, *Journal of the Royal Statistical Society, Series C, Applied Statistics*, 58, pp. 467–484.

Goldman, N. and Yang, Z. (1994). A codon-based model of nucleotide substitution for protein-coding DNA sequences, *Molecular Biology and Evolution*, 11, pp. 725–736.

Goldstein, R. A. and Pollock, D. D. (2017). Sequence entropy of folding and the absolute rate of amino acid substitutions, *Nature Ecology and Evolution*, 1, pp. 1923–1930.

Gonnet, G. H., Cohen, M. A. and Benner, S. A. (1992). Exhaustive matching of the entire protein sequence database, *Science*, 256, pp. 1443–1445.

Grantham, R. (1974). Amino acid difference formula to help explain protein evolution, *Science*, 185, pp. 862–864.

Gray, R. M. (1984). Vector quantization, *IEEE ASSP Magazine*, 1, pp. 4–29.

Greenfield, L. M., Hill, P. W., Paterson, E., Baggs, E. M. and Jones, D. L. (2020). Do plants use root-derived proteases to promote the uptake of soil organic nitrogen? *Plant and Soil*, 456, pp. 355–367.

Griswold, A. R., Cifani, P., Rao, S. D., Axelrod, A. J., Miele, M. M., Hendrickson, R. C., Kentsis, A. and Bachovchin, D. A. (2019). A chemical strategy for protease substrate profiling, *Cell Chemical Biology*, 26, pp. 901–907.

Gumaei, A., Ismail, W. N., Hassan, M. R., Hassan, M., Hassan, M. M., Mohamed, E., Aleaiwi, A. and Fortino, G. (2022). A decision-level fusion method for COVID-19 patient health prediction, *Big Data Research*, 27, pp. 100287.

Guo, H. H., Choe, J. and Loeb, L. A. (2004). Protein tolerance to random amino acid change, *Proceedings of the National Academy of Sciences of the United States of America*, 101, pp. 9205–9210.

Guzzo, A. V. (1965). The influence of amino-acid sequence on protein structure, *Biophysical Journal*, 5, pp. 809–812.

Hall, D. L. and Llinas, J. (1997). An introduction to multisensor data fusion, *Proceedings of the IEEE*, 85, pp. 6–23.

Hanley, J. A. and McNeil, B. J. (1982). The Meaning and Use of the Area under a Receiver Operating Characteristic (ROC) Curve, *Radiology*, 143, pp. 29–36.

Hanson, M. A. and Marzluf, G. A. (1975). Control of the synthesis of a single enzyme by multiple regulatory circuits in Neurospora crassa, *Proceedings of the National Academy of Sciences of the United States of America*, 72, pp. 1240–1244.

Hartman, E., Keeler, J. D. and Kowalski, J. M. (1990). Layered neural networks with Gaussian hidden units as universal approximations, *Neural Computation*, 2, pp. 210–215.

Hasan, A., Paray, B. A., Hussain, A., Qadir, F. A., Attar, F., Aziz, F. M., Sharifi, M., Derakhshankhah, H., Rasti, B., Mehrabi, M., Shahpasand, K., Saboury, A. A. and Falahati, M. (2021). A review on the cleavage priming of the spike protein on coronavirus by angiotensin-converting enzyme-2 and furin, *Journal of Biomolecular Structure* and *Dynamics*, 39, pp. 3025–3033.

Hastie, T. J., Tibshirani, R. and Friedman, J. (2001). *The Elements of Statistical Learning*, Springer.

Hearst, M. A., Dumais, S. T., Osuna, E., Platt, J. and Scholkopf, B. (1998). Support vector machines, *IEEE Intelligent Systems and their Applications*, 13, pp. 18–28.

Hein, J. E. (2021). Machine learning made easy for optimizing chemical reactions, *Nature*, 590, pp. 40–41.

Henikoff, S. and Henikoff, J. G. (1992). Amino acid substitution matrices from protein blocks, *Proceedings of the National Academy of Sciences of the United States of America*, 89, pp. 10915–10919.

Hert, J., Willett, P., Wilton, D. J., Acklin, P., Azzaoui, K., Jacoby, E. and Schuffenhauer, A. (2006). New methods for ligand-based virtual screening: use of data fusion and machine learning to enhance the effectiveness of similarity searching, *Journal of Chemical Information and Modeling*, 46, pp. 462–470.

Hilton, S. K. and Bloom, J. D. (2018). Modeling site-specific amino-acid preferences deepens phylogenetic estimates of viral sequence divergence, *Viral evolution*, 4, pp. 33.

Ho, T. K. (1998). The random subspace method for constructing decision forests, *IEEE Transactions on Pattern Analysis and Machine Intelligence*, 20, pp. 832–844.

Hoerl, A. E. (1962). Application of ridge analysis to regression problems, *Chemical Engineering Progress*, 58, pp. 54–59.

Hoerl, A. E. and Kennard, R. W. (1970a). Ridge regression: biased estimation for nonorthogonal problems, *Technometrics*, 12, pp. 55–67.

Hoerl, A. E. and Kennard, R. W. (1970b). Ridge regression: applications to nonorthogonal problems, *Technometrics*, 12, pp. 69–82.

Huang, J. and Brutlag, D. (2001). The emotif database, *Nucleic Acids Research*, 29, pp. 202–204.

Huang, S. Y., Yeh, Y. R. and Eguchi, S. (2009). Robust kernel principal component analysis, *Neural Computation*, 21, pp. 3179–3213.

Huelsenbeck, J. P., Joyce, P., Lakner, C. and Ronquist, F. (2008). Bayesian analysis of amino acid substitution models, *Philosophical Transactions of the Royal Society B*, 363, pp. 3941–3953.

Jesneck, J. L., Nolte, L. W., Baker, J. A., Floyd, C. E. and Lo, J. Y. (2006). Optimized approach to decision fusion of heterogeneous data for breast cancer diagnosis, *Medical Physics*, 33, pp. 2945–2954.

Johnson, M. S. and Overington, J. P. (1993). A structural basis for sequence comparisons: An evaluation of scoring methodologies, *Journal of Molecular Biology*, 233, pp. 716–738.

Jombart, T., Devillard, S. and Balloux, F. (2010). Discriminant analysis of principal components: a new method for the analysis of genetically structured populations, *BMC Genetics*, 11, pp. 94.

Jones, D. T., Taylor, W. R. and Thornton, J. M. (1992). The rapid generation of mutation data matrices from protein sequences, *Computer Applications in the Biosciences*, 8, pp. 275–282.

Jukes, T. H. and Cantor, C. R. (1969). Evolution of protein molecules, *Mammalian Protein Metabolism*, Academic Press, New York, pp. 21–132.

Kaiser, A. and Agostinelli, E. (2022). Hypusinated EIF5A as a feasible drug target for Advanced Medicinal Therapies in the treatment of pathogenic parasites and therapy-resistant tumors, *Amino Acids*, 54, pp. 501–511.

Kann, M., Qian, B. and Goldstein, R. A. (2000). Optimization of a new score function for the detection of remote homologs, *Proteins*, 41, pp. 498–503.

Karatsuba, A. A. and Ofman, Y. P. (1963). Multiplication of multidigit numbers on automata, *Soviet Physics Doklady*, 7, pp. 595–596.

Kasturi, J. and Acharya, R. (2005). Clustering of diverse genomic data using information fusion, *Bioinformatics*, 21, pp. 423–429.

Kelleci, C. F. and Karaduman, G. (2023). In silico QSAR modeling to predict the safe use of antibiotics during pregnancy, *Drug and Chemical Toxicology*, 46, pp. 962–971.

Kende, J., Bonomi, M., Temmam, S., Regnault, B., Perot, P., Eloit, M. and Bigot, T. (2023). Virus pop-expanding viral databases by protein sequence simulation, *Viruses*, 15, pp. 1227.

Keul, F., Hess, M., Goesele, M. and Hamacher, K. (2017). PFASUM: a substitution matrix from Pfam structural alignments, *BMC Bioinformatics*, 18, pp. 293.

Khare, S., *et al.* (2021). GISAID's Role in Pandemic Response, *China CDC Weekly*, 3, pp. 1049–1051.

Kilmer, M. E., Hansen, P. C. and Espanol, M. I. (2007). A projection-based approach to general-form Tikhonov regularization, *SIAM Journal of Scientific Computing*, 29, pp. 315–330.

Kimura, M. (1980). A simple method for estimating evolutionary rates of base substitutions through comparative studies of nucleotide sequences, *Journal of Molecular Evolution*, 16, pp. 111–120.

Konc, J. and Janezic, D. (2007) Protein-protein binding-sites prediction by protein surface structure conservation, *Journal of Chemical Information and Modeling*, 47, pp. 940–944.

Kohonen, T. (1989). *Self-Organization and Associative Memory*, 3rd ed., Springer-Verlag, Berlin.

Kowalski, P. A. and Kusy, M. (2018). Sensitivity analysis for probabilistic neural network structure reduction, *IEEE Transactions on Neural Network Learning Systems*, 29, pp. 1919–1932.

Kuang, R., Ie, E., Wang, K., Wang, K., Siddiqi, M., Freund, Y. and Leslie, C. (2005). Profile-based string kernels for remote homology detection and motif extraction, *Journal of bioinformatics and Computational Biology*, 3, pp. 527–550.

Kuksa, P. P., Huang, P. H. and Pavlovic, V. (2008). Scalable algorithms for string kernels with inexact matching, *Proceedings of Advances in Neural Information Processing Systems*, pp. 881–888.

Lau, K. W., Yin, H. and Hurbard, S. (2006). Kernel self-organising maps for classification, *Neurocomputing*, 69, pp. 2033–2040.

Laua, K. W., Yin, H. and Hubbard, S. (2006). Kernel self-organising maps for classification, *Neurocomputing*, 69, pp. 2033–2040.

LeCun, Y., Bengio, Y. and Hinton, G. (2015). Deep learning, *Nature*, 521, pp. 436–444.

Leslie, C. and Kuang, R. (2003). Fast kernels for inexact string matching, Computational Learning Theory and Kernel Machines, *16th Annual Conference on Computational Learning Theory and 7th Kernel Workshop*, Washington, DC, USA.

Leslie, C. S., Eskin, E., Cohen, A., Weston, J. and Noble, W. S. (2004). Mismatch string kernels for discriminative protein classification, *Bioinformatics*, 20, pp. 467–476.

Leslie, C. and Kuang, R. (2004). Fast string kernels using inexact matching for protein sequences, *Journal of Machine Learning Research*, 5, pp. 1435–1455.

Leslie, C., Eskin, E. and Noble, W. S. (2002). The spectrum kernel: a string kernel for SVM protein classification, *Proceedings of Pacific Biocomputing Symposium*, 7, pp. 566–575.

Levin, J. M., Robson, B. and Garnier, J. (1986). An algorithm for secondary structure determination in proteins based on sequence similarity, *FEBS Letters*, 205, pp. 303–308.

Li, Y., Gong, S. and Liddell, H. (2003). Recognising trajectories of facial identities using kernel discriminant analysis, *Image and Vision Computing*, 21, pp. 1077–1086.

Liang, G., Yang, L., Chen, Z., Mei, H., Shu, M. and Li, Z. (2009). A set of new amino acid descriptors applied in prediction of MHC class I binding peptides, *European Journal of Medicinal Chemistry*, 44, pp. 1144–1154.

Lipman, D. J., Altschul, S. F. and Kececioglu, J. D. (1989). A tool for multiple sequence alignment, *Proceedings of the National Academy of Sciences*, 86, pp. 4412–4415.

Liu, X. (2023). SimpleMKKM: Simple multiple kernel k-means, *IEEE Transactions on Pattern Analysis and Machine Intelligence*, 45, pp. 5174–5186.

Lodhi, H., Saunders, C., Shawe-Taylor, J., Cristianini, N. and Watkins, C. (2002). Text classification using string kernels, *Journal of Machine Learning Research*, 2, pp. 419–444.

Lu, Y., Sundararajan, N. and Saratchandran, P. (1997). A sequential learning scheme for function approximation using minimal radial basis function neural networks, *Neural Computation*, 9, pp. 461–478.

Luo, M., Capina, R., Daniuk, C., Tuff, J., Peters, H., Kimani, M., Wachihi, C., Kimani, J., Ball, T. B. and Plummer, F. A. (2013). Immunogenicity of sequences around HIV-1 protease cleavage sites: potential targets and population coverage analysis for a HIV vaccine targeting protease cleavage sites, *Vaccine*, 31, pp. 3000–3008.

Luo, S. Y., Araya, L. E. and Julien, O. (2019). Protease substrate identification using N-terminomics, *ACS Chemical Biology*, 11, pp. 2361–2371.

Luong, T. V., Shlezinger, N., Xu, C., Hoang, T. M., Eldar, Y. C. and Hanzo, L. (2022). Deep learning based successive interference cancellation for the non-orthogonal downlink, *IEEE Transactions on Vehicular Technology*, 71, pp. 11876–11888.

Lustosa, D. M. and Milo, A. (2023). Machine learning classifies catalytic-reaction mechanisms, *Nature*, 613, pp. 635–636.

Lyle, P. and Trimble, S. Y. (1991). Gram-Schmidt orthogonalization by Gauss elimination, *The American Mathematical Monthly*, 98, pp. 544–549.

MacDonald, D. and Fyfe, C. (2000). The kernel self-organizing map, *The 4th International Conference on Knowledge-Based Intelligent Engineering Systems and Allied Technologies*, Brighton, UK.

Macfarlane, M. G. (1936). Phosphorylation in living yeast, *Biochemical Journal*, 30, pp. 1369–1379.

MacQueen, J. B. (1967). Some methods for classification and analysis of multivariate observations, *Proceedings of 5-th Berkeley Symposium on Mathematical Statistics and Probability*, Berkeley, University of California Press, 1, pp. 281–297.

Matthews, B. W. (1975). Comparison of the predicted and observed secondary structure of T4 phage lysozyme, *Biochimica et Biophysica Acta (BBA) — Protein Structure*, 405, pp. 442–451.

McDonald, G. (2009). Ridge regression, *Wiley Interdisciplinary Reviews: Computational Statistics*, 1, pp. 93–100.

Meng, F., Liang, Z., Zhao, K. and Luo, C. (2021). Drug design targeting active posttranslational modification protein isoforms, *Medicinal Research Reviews*, 41, pp. 1701–1750.

Meng, X. L. and van Dyk, D. (1997). The EM algorithm — an old folk-song sung to a fast new tune, *Journal of Royal Statistical Society B*, 59, pp. 511–567.

Metsis, V., Huang, H., Andronesi, O. C., Makedon, F. and Tzika, A. (2012). Heterogeneous data fusion for brain tumor classification, *Oncology Reports*, 28, pp. 1413–1416.

Mika, S., Rätsch, G., Weston, J., Schölkopf, B. and Müller, K. R. (1999). Fisher discriminant analysis with kernels, *Neural Networks for Signal Processing*, IX, pp. 41–48.

Meyer, C. D. (2000). *Matrix Analysis and Applied Linear Algebra*, SIAM, USA.

Mirsky, A., Kazandjian, L. and Anisimova, M. (2015). Antibody-specific model of amino acid substitution for immunological inferences from alignments of antibody sequences, *Molecular Biology and Evolution*, 32, pp. 806–819.

Mitchell, T. M. (1997). *Machine Learning*, McGraw-Hill International Editions Computer Science Series.

Miyata, T., Miyazawa, S. and Yasunaga, T. (1979). Two types of amino acid substitutions in protein evolution, *Journal of Molecular Evolution*, 12, pp. 219–236.

Moelleken, J., Endesfelder, M., Gassner, C., Lingke, S., Tomaschek, S., Tyshchuk, O., Lorenz, S., Reiff, U. and Molhojm M. (2017). GingisKHAN^{TM} protease cleavage allows a high-throughput antibody to Fab conversion enabling direct functional assessment during lead identification of human monoclonal and bispecific IgG1 antibodies, *MAbs*, 9, pp. 1076–1087.

Mohana R., J. (1987). New scoring matrix for amino acid residue exchanges based on residue characteristic physical parameters, *International Journal of Peptide Protein Research*, 29, pp. 276–281.

Moravcikova, N., Kasarda, R., Zidek, R., McEwan, J. C., Brauning, R., Landete-Castillejos, T., Chonco, L., Ciberej, J. and Pokoradi, J. (2023). Traces of human-mediated selection in the gene pool of red deer populations, *Animals*, 13, pp. 2525.

Mousavi, S. L. and Sajjadi, S. M. (2023). Predicting rejection of emerging contaminants through RO membrane filtration based on ANN-QSAR modeling approach: trends in molecular descriptors and structures towards rejections, *RSC Advances*, 13, pp. 23754–23771.

Muller, T. and Vingron, M. (2000). Modeling amino acid replacement, *Journal of Computational Biology*, 7, pp. 761–776.

Muller, T., Rahmann, S. and Rehmsmeier, M. (2001). Non-symmetric score matrices and the detection of homologous transmembrane proteins, *Bioinformatics*, 17, pp. S182–S189.

Muller, T., Spang, R. and Vingron, M. (2002). Estimating amino acid substitution models: a comparison of Dayhoff's estimator, the resolvent approach and a maximum likelihood method, *Molecular Biology and Evolution*, 19, pp. 8–13.

Musavi, M. T., Ahmed, W., Chan, K. H., Faris, K. B. and Hummels, D. M. (1992). On the training of radial basis function classifiers, *Neural Networks*, 5, pp. 595–603.

Muse, S. V. (1996). Estimating synonymous and nonsynonymous substitution rates, *Molecular Biology and Evolution*, 13, pp. 105–114.

Nadeem, M. W., Goh, H. G., Ponnusamy, V., Andonovic, I., Khan, M. A. and Hussain, M. (2021). A fusion-based machine learning approach for the prediction of the onset of diabetes, *Healthcare*, 9, pp. 2–16.

Naor, D., Fischer, D., Jernigan, R. L., Wolfson, H. J. and Nussinov, R. (1996). Amino acid pair interchanges at spatially conserved sites, *Journal of Molecular Biology*, 256, pp. 924–938.

Narayanan, A., Wu, C. and Yang, Z. R. (2002). Mining viral protease data to extract cleavage knowledge, *Bioinformatics*, 18, pp. s5–s13.

Needleman, S. B. and Wunsch, C. D. (1970). A general method applicable to the search for similarities in the amino acid sequence of two proteins, *Journal of Molecular Biology*, 48, pp. 443–453.

Nei, M. and Gojobori, T. (1986). Simple methods for estimating the numbers of synonymous and nonsynonymous nucleotide substitutions, *Molecular Biology and Evolution*, 3, pp. 418–426.

Nevill-Manning, C., Wu, T. and Brutlag, D. (1998). Highly specific protein sequence motifs for genome analysis, *Proceedings of the National Academy of Sciences of the United States of America*, 95, pp. 5865–5871.

Nivethitha, S., Gauthama, R., Kalpana, V., Kannan, K. and Shankar, S. (2018). An improved robust heteroscedastic probabilistic neural network based trust prediction approach for cloud service selection, *Neural Networks*, 108, pp. 339–354.

Overington, J. P., Johnson, M. S., Sali, A. and Blundell, T. L. (1990). Tertiary structural constraints on protein evolutionary diversity, *Proceedings of the Royal Society B: Biological Sciences*, 241, pp. 132–145.

Palme, J., Hochreiter, S. and Bodenhofer, U. (2015). KeBABS: an R package for kernel-based analysis of biological sequences, *Bioinformatics*, 31, pp. 2574–2576.

Palzkill, T. and Botstein, D. (1992). Probing beta-lactamase structure and function using random replacement mutagenesis, *Proteins*, 14, pp. 29–44.

Papadimitriou, S. and Likothanassis, S. D. (2004). Kernel-based self-organized maps trained with supervised bias for gene expression data analysis, *Journal of Bioinformatics and Computational Biology*, 1, pp. 647–680.

Papoulis, A. and Pillai, S. U. (2002). *Probability, Random Variables and Stochastic Processes*, McGraw-Hill.

Park, J. and Sandberg, I. W. (1991). Universal approximation using radial-basis-function networks, *Neural Computation*, 3, pp. 246–257.

Parzen, E. (1962). Extraction and detection problems and reproducing kernel Hilbert spaces, *Journal of the Society for Industrial and Applied Mathematics, Series A, On control*, 1, pp. 35–63.

Piryonesi, S. M. and El-Diraby, T. E. (2020). Data analytics in asset management: cost-effective prediction of the pavement condition index, *Journal of Infrastructure Systems*, 26, pp. 1.

Prlic, A., Domingues, F. S. and Sippl, M. J. (2000). Structure-derived substitution matrices for alignment of distantly related sequences, *Protein Engineering*, 13, pp. 545–550.

Punekar, N. S. (2018). *Enzymes: Catalysis, Kinetics and Mechanisms*, Springer Nature Singapore Pte Ltd.

Quinlan, J. R. (1986). Induction of decision trees, *Machine Learning*, 1, pp. 81–106.

Quinlan, J. R. (1993). *C4.5: Programs for Machine Learning*, Morgan Kaufmann Publishers.

Reardon, S. (2019). How machine learning could keep dangerous DNA out of terrorists' hands, *Nature*, 566, pp. 19.

Rimington, C. (1927). Phosphorylation of proteins, *Biochemical Journal*, 21, pp. 272–281.

Risler, J., Delorme, M., Delacroix, H. and Henaut, A. (1988). Amino acid substitutions in structurally related proteins a pattern recognition approach: determination of a new and efficient scoring matrix, *Journal of Molecular Biology*, 204, pp. 1019–1029.

Rosenblatt, M. (1956). Remarks on some nonparametric estimates of a density function, *The Annals of Mathematical Statistics*, 27, pp. 832–837.

Rubinstein, N. D., Feldstein, T., Shenkar, N., Botero-Castro, F., Griggio, F., Mastrototaro, F., Delsuc, F., Douzery, E. J., Gissi, C. and Huchon, D. (2013). Deep sequencing of mixed total DNA without barcodes allows efficient assembly of highly plastic ascidian mitochondrial genomes, *Genome Biology and Evolution*, 5, pp. 1185–1199.

Russell, R. B., Saqi, M. A. S., Sayle, R. A., Bates, P. A. and Sternberg, M. J. E. (1997). Recognition of analogous and homologous protein folds. Analysis of sequence and structure conservation, *Journal of Molecular Biology*, 269, pp. 423–439.

Sagi, O. and Rokach, L. (2021). Approximating XGBoost with an interpretable decision tree, *Information Sciences*, 572, pp. 522–542.

Sammon, J. W. (1969). A nonlinear mapping for data structure analysis, *IEEE Transactions on Computers*, 18, pp. 401–402.

Sammut, S. J., Crispin-Ortuzar, M., Chin, S. F., Provenzano, E., Bardwell, H. A., Ma, W., Cope, W., Dariush, A., Dawson, S. J., Abraham, J. E., Dunn, J., Hiller, L., Thomas, J., Cameron, D. A., Bartlett, J. M. S., Hayward, L., Pharoah, P. D., Markowetz, F., Rueda, O. M., Earl, H. M. and Caldas, C. (2022). Multi-omic machine learning predictor of breast cancer therapy response, *Nature*, 601, pp. 623–629.

Santosa, F. and Symes, W. W. (1986). Linear inversion of band-limited reflection seismograms, *SIAM Journal on Scientific and Statistical Computing*, 7, pp. 1307–1330.

Satish, D. S. and Sekhar, C. (2006). Kernel based clustering and vector quantization for speech segmentation, *The International Joint Conference on Neural Networks*, Vancouver, BC, Canada, 16–21 July 2006.

Sato, A. and Yamada, K. (1995). Generalized learning vector quantization, *Advances in Neural Information Processing Systems 8*.

Savage, N. (2017). Machine learning: calculating disease, *Nature*, 550, pp. S115–S117.

Savchenko, A. (2020). Probabilistic neural network with complex exponential activation functions in image recognition, *IEEE Transactions on Neural Network Learning Systems*, 31, pp. 651–660.

Schiffer, M. and Edmundson, A. B. (1968). Correlation of amino acid sequence and conformation in tobacco mosaic virus, *Biophysical Journal*, 8, pp. 29–33.

Schmidhuber, J. (2015). Deep learning in neural networks: an overview, *Neural Networks*, 61, pp. 85–117.

Scholkopf, B., Smola, A. and Muller, K. R. (1998). Nonlinear component analysis as a kernel eigenvalue problem, *Neural Computation*, 10, pp. 1299–1319.

Scaranto, B. M. S., Ribolli, J., Vieira, G. C., Ferreira, J. P. R., de Miranda Gomes, C. H. A. and de Melo, C. M. R. (2023). Genetic structure and diversity in wild and cultivated populations of the mangrove oyster crassostrea gasar from southern Brazil, *Marine Biotechnology*, 25, pp. 548–556.

Sellers, P. H. (1974). On the theory and computation of evolutionary distances, *SIAM Journal on Applied Mathematics*, 26, pp. 787–793.

Shapiro, L. G. and Haralick, R. M. (1982). Organization of relational models for scene analysis, *IEEE Transactions on Pattern Analysis and Machine Intelligence*, 4, pp. 595–602.

Shu, Y. and McCauley, J. (2017). GISAID: from vision to reality, *EuroSurveillance*, 22, pp. 13.

Singh, S. P. and Jagg, M. (2020). Model fusion via optimal transport, *The 34th Conference on Neural Information Processing Systems*, Vancouver, Canada.

Smith, T. F. and Waterman, M. S. (1981). Identification of common molecular subsequences, *Journal of Molecular Biology*, 147, pp. 195–197.

Soklakov, A. N. (2002). Occam's Razor as a formal basis for a physical theory, *Foundations of Physics Letters*, 15, pp. 107–135.

Soong, F., Rosenberg, A., Rabiner, L. and Juang, B. (1985). A vector quantization approach to speaker recognition, *IEEE Proceedings International Conference on Acoustics, Speech and Signal Processing*, 1, pp. 387–339.

Specht, D. F. (1990). Probabilistic neural networks, *Neural Networks*, 3, 109–118.

Stamenkovic, M., Steinwall, E., Nilsson, A. K. and Wulff, A. (2020). Fatty acids as chemotaxonomic and ecophysiological traits in green microalgae (desmids, Zygnematophyceae, Streptophyta): a discriminant analysis approach, *Phytochemistry*, 170, pp. 112200.

Taillon-Miller, P. A. and Shreffler, D. C. (1988). Structural basis for the C4d.1/C4d.2 serologic allotypes of murine complement component C4, *Journal of Immunology*, 141, pp. 2382–2387.

Taylor, D. J., Green, N. P. O. and Stout, G. W. (1997). *Biological Science* 1 & 2, 3rd edition, Cambridge University Press.

Thomson, R., Hodgman, T. C., Yang, Z. R. and Doyle, A. K. (2003). Characterizing proteolytic cleavage site activity using bio-basis function neural networks, *Bioinformatics*, 19, pp. 1741–1747.

Thompson, J. D., Plewniak, F. and Poch, O. (1999). A comprehensive comparison of multiple sequence alignment programs, *Nucleic Acids Research*, 27, pp. 2682–2690.

Tibshirani, R. (1996). Regression shrinkage and selection via the lasso, *Journal of the Royal Statistical Society. Series B (methodological)*, 58, pp. 267 88.

Tibshirani, R. (1997). The lasso method for variable selection in the Cox model, *Statistics in Medicine*, 16, pp. 385–395.

Tipping, M. E. (2001). Sparse Bayesian learning and the relevance vector machine, *Journal of Machine Learning Research*, 1, pp. 211–244.

Tong, J., Liu, S., Zhou, P., Wu, B. and Li, Z. (2008). A novel descriptor of amino acids and its application in peptide QSAR, *Journal of Theoretical Biology*, 253, pp. 90-97.

Trivedi, R. and Nagarajaram, H. A. (2019). Amino acid substitution scoring matrices specific to intrinsically disordered regions in proteins, *Scientific Reports*, 9, pp. 16380.

Trudgian, D. and Yang, Z. R. (2007). Substitution matrix optimisation for peptide classification, *Lecture Notes in computer Science*, 4447, pp. 291–300.

Tseng, Y. Y. and Liang, J. (2006). Estimation of amino acid residue substitution rates at local spatial regions and application in protein function inference: a Bayesian Monte Carlo approach, *Molecular Biology and Evolution*, 23, pp. 421–436.

Tzortzis, G. F. and Likas, A. C. (2009). The global kernel k-means algorithm for clustering in feature space, *IEEE Transactions on Neural Networks*, 20, pp. 1181–1194.

van Westen, G. J., Swier, R. F., Cortes-Ciriano, I., Wegner, J. K., Overington, J. P., Ijzerman, A. P., van Vlijmen, H. W. and Bender, A. (2013). Benchmarking of protein descriptor sets in proteochemometric modeling (part 2): modeling performance of 13 amino acid descriptor sets, *Journal of Cheminformatics*, 5, pp. 42.

Vapnik, V. (1995). *The Nature of Statistical Learning Theory*, Springer-Verlag, New York.

Venables, W. N. and Ripley, B. D. (2002). *Modern Applied Statistics with S-PLUS*, 4th edition, Springer: Berlin/Heidelberg, Germany.

Venkatesh, S. and Gopal, S. (2011). Robust Heteroscedastic Probabilistic Neural Network for multiple source partial discharge pattern recognition — significance of outliers on classification capability, *Expert Systems with Applications*, 38, pp. 11501–11514.

Villa, N. and Rossi, F. (2007). A comparison between dissimilarity SOM and kernel SOM for clustering the vertices of a graph, *The 6th International Workshop on Self-Organizing Maps*, Germany.

Vishwanathan, S. V. N. and Smola, A. (2002). Fast kernels for string and tree matching, *Neural Information Processing Systems*, 15, pp. 569–576.

Vogt, G., Etzold, T. and Argos, P. (1995). An assessment of amino acid exchange matrices in aligning protein sequences: the twilight zone revisited, *Journal of Molecular Biology*, 249, pp. 816–831.

von der Malsburg, C. (1973). Self-organization of orientation sensitive cells in the striate cortex, *Kybernetik*, 14, pp. 85–100.

Ward, C. and Kincaid, D. (2009). *Linear Algebra: Theory and Applications*, Sudbury, Ma: Jones and Bartlett.

Wagih, O. (2017). ggseqlogo: a versatile R package for drawing sequence logos, *Bioinformatics*, 33, pp. 3645–3647.

Wang, F., Zhao, H., Yu, C., Tang, J., Wu, W. and Yang, Q. (2020). Determination of the geographical origin of maize (Zea mays L.) using mineral element fingerprints, *Journal of Science of Food Agriculture*, 100, pp. 1294–1300.

Webb, A. (2002). *Statistical Pattern Recognition*, 2nd Edition, John Wiley & Sons Ltd.

Weber, I. T., Wang, Y. F. and Harrison, R. W. (2021). HIV protease: historical perspective and current research, *Viruses*, 13, pp. 839.

Weisberg, S. (2005). *Applied Linear Regression*, John Wiley & Sons, New York.

Wernersson, R., Rapacki, K., Staerfeldt, H. H., Sackett, P. W. and Molgaard, A. (2006). FeatureMap3D–a tool to map protein features and sequence conservation onto homologous structures in the PDB, *Nucleic Acids Research*, 34, pp. W84–W88.

Wiggert, W. P. and Werkman, C. H. (1938). Phosphorylation by the living bacterial cell, *Biochemical Journal*, 32, pp. 101–107.

Wisniowska, B., Mendyk, A., Szlek, J., Kolaczkowski, M. and Polak, S. (2015). Enhanced QSAR models for drug-triggered inhibition of the main cardiac ion currents, *Journal of Applied Toxicology*, 35, pp. 1030–1039.

Wong, D. and Yip, S. (2018). Machine learning classifies cancer, *Nature*, 555, pp. 446–447.

Wong, T. S., Roccatano, D., Zacharias, M. and Schwaneberg, U. (2006). A statistical analysis of random mutagenesis methods used for directed protein evolution, *Journal of Molecular Biology*, 355, pp. 858–871.

Xie, R. K., Long, H. X., Cheng, X. M., Wang, Y. Q., Lin, Y., Yang, Y., Zhu, B. and Lin, Z. H. (2010). Quantitative sequence–kinetics relationship in antigen–antiboby interaction kinetics based on a set of descriptors, *Chemical Biology & Drug Design*, 76, pp. 345–349.

Yang, L., Shu, M., Ma, K., Mei, H., Jiang, Y. and Li, Z. (2010), ST-scale as a novel amino acid descriptor and its application in QSAM of peptides and analogues, *Amino Acids*, 38, pp. 805–816.

Yang, Z., and Nielsen, R. (2000). Estimating synonymous and nonsynonymous substitution rates under realistic evolutionary models, *Molecular Biology and Evolution*, 17, pp. 32–43.

Yang, Z. R. (2005a). Mining SARS-CoV protease cleavage data using non-orthogonal decision trees: a novel method for decisive template selection, *Bioinformatics*, 21, pp. 2644–2650.

Yang, Z. R. (2005b). Prediction of caspase cleavage sites using Bayesian bio-basis function neural networks, *Bioinformatics*, 21, pp. 1831–1837.

Yang, Z. R. (2005c). Orthogonal kernel machine for the prediction of functional sites in proteins, *IEEE Transactions on Systems and Cybernetics B*, 35, pp. 100–106.

Yang, Z. R. (2006). Predicting hepatitis C virus protease cleavage sites using generalized linear indicator regression models, *IEEE Transactions on Biomedical Engineering*, 53, pp. 2119–2123.

Yang, Z. R. (2022a). *Biological Pattern Discovery with R, Machine Learning Approaches*, World Scientific Publisher.

Yang, Z. R. (2022b). In silico prediction of severe acute respiratory syndrome coronavirus 2 main protease cleavage sites, *PROTEINS: Structure, Function, and Bioinformatics*, 90, pp. 791–801.

Yang, Z. R. and Berry, E. A. (2004). Reduced bio-basis function neural networks for protease cleavage site prediction, *Journal of Bioinformatics and Computational Biology*, 2, pp. 511–531.

Yang, Z. R., Dalby, A. R. and Qiu, J. (2004). Mining HIV protease cleavage data using genetic programming with a sum-product function, *Bioinformatics*, 20, pp. 3398–3405.

Yang, Z. R. and Chen, S. (1998). Robust maximum likelihood training of heteroscedastic probabilistic neural networks, *Neural Networks*, 11, pp. 739–741.

Yang, Z. R. and Chou, K. C. (2004a). Bio-support vector machines for computational proteomics, *Bioinformatics*, 20, pp. 735–741.

Yang, Z. R. and Chou, K. C. (2004b). Predicting the linkage sites in glycoproteins using bio-basis function neural network, *Bioinformatics*, 20, pp. 903–908.

Yang, Z. R., Dry, J., Thomson, R. and Hodgman, T. C. (2006). A bio-basis function neural network for protein peptide cleavage activity characterisation, *Neural Networks*, 19, pp. 401–407.

Yang, Z. R. and Hamer, R. (2007). Bio-basis function neural networks in protein data mining, *Current Pharmaceutical Design*, 13, pp. 1403–1413.

Yang, Z. R. and Young, N. (2005). Bio-kernel self-organizing map for HIV drug resistance classification, *Advances in Natural Computation*, 3610, pp.179–186.

Yang, Z. R. and Thomson, R. (2005). Bio-basis function neural network for prediction of protease cleavage sites in proteins, *IEEE Transactions on Neural Network*, 16, pp. 263–274.

Yang, Z. R., Thomson, R., Hodgman, T. C., Dry, J., Doyle, A. K., Narayanan, A. and Wu, X. (2003). Searching for discrimination rules in protease proteolytic cleavage activity using genetic programming with a min-max scoring function, *Biosystems*, 72, pp. 159–176.

Yang, Z. R., Thomson, R., McNeil, P. and Esnouf, P. M. (2005). RONN: the bio-basis function neural network technique applied to the detection of natively disordered regions in proteins. *Bioinformatics*, 21, pp. 3369–3376.

Yang, Z. R., Zwolinski, M., Chalk, C. D. and Williams, A. C. (2000). Applying a robust heteroscedastic probabilistic neural network to analog fault detection and classification, *IEEE Transactions on Computer-Aided Design of Integrated Circuits and Systems*, 10, pp. 142–151.

Yao, P., Zhu, Q. and Zhao, R. (2022). Gaussian mixture model and self-organizing map neural-network-based coverage for target search in curve-shape area, *IEEE Transactions on Systems, Man, and Cybernetics*, 52, pp. 3971–3983.

Yin, H. (2006). On the equivalence between kernel self-organising maps and self-organising mixture density networks, *Neural Networks*, 19, pp. 780–784.

Yin, H. and Allinson, N. (2001). Self-organising mixture networks for probability density estimation, *IEEE Transactions on Neural Networks*, 12, pp. 405–411.

van Hulle, M. (2002). Kernel-based topographic map formation achieved with an information- theoretic approach. *Neural Networks*, 15, pp. 1029–1039.

Yousefzadeh, H., Raeisi, S., Esmailzadeh, O., Jalali, G., Nasiri, M., Walas, L. and Kozlowski, G. (2021). Genetic diversity and structure of rear edge populations of *sorbus aucuparia* (Rosaceae) in the Hyrcanian forest, *Plants*, 10, pp. 1471.

Yu, D., Qi, Y., Xu, Y. and Yang, J. (2023). Kernel-SOM based visualization of financial time series forecasting, *The First International Conference on Innovative Computing, Information and Control*, China.

Yu, Q., Jalaludin, A., Han, H., Chen, M., Sammons, R. D. and Powles, S. B. (2015). Evolution of a double amino acid substitution in the 5-enolpyruvylshikimate-3-phosphate synthase in Eleusine indica conferring high-level glyphosate resistance, *Plant Physiology*, 167, pp. 1440–1447.

Zhao, J., Kardashliev, T., Joelle, R. A., Bocola, M. and Schwaneberg, U. (2014). Lessons from diversity of directed evolution experiments by an analysis of 3,000 mutations, *Biotechnology Bioengineering*, 111, pp. 2380–2389.

Index

www.ingramcontent.com/pod-product-compliance
Lightning Source LLC
Chambersburg PA
CBHW050551190326
41458CB00007B/1995